SCIENCE AND SCRIPTURE

How Science Deepens One's
Understanding of Biblical Passages

Other World Scientific Titles on the Subject

Cosmic Roots: The Conflict Between Science and Religion and How it Led to the Secular Age
by Ira Mark Egdall
ISBN: 978-981-125-138-2
ISBN: 978-981-125-247-1 (pbk)

God or Science?: Is Science Denying God?
by Antonino Del Popolo
ISBN: 978-981-125-872-5
ISBN: 978-981-126-558-7 (pbk)

Reverse Engineering God: Irreligious Answers to Fundamental Questions
by Michael Rothschild
ISBN: 978-981-123-090-5

SCIENCE AND SCRIPTURE

How Science Deepens One's Understanding of Biblical Passages

Nathan Aviezer

Bar-Ilan University, Israel

World Scientific

NEW JERSEY · LONDON · SINGAPORE · BEIJING · SHANGHAI · HONG KONG · TAIPEI · CHENNAI · TOKYO

Published by

World Scientific Publishing Co. Pte. Ltd.

5 Toh Tuck Link, Singapore 596224

USA office: 27 Warren Street, Suite 401-402, Hackensack, NJ 07601

UK office: 57 Shelton Street, Covent Garden, London WC2H 9HE

Library of Congress Control Number: 2022028216

British Library Cataloguing-in-Publication Data
A catalogue record for this book is available from the British Library.

SCIENCE AND SCRIPTURE
How Science Deepens One's Understanding of Biblical Passages

ISBN 978-981-126-430-6 (hardcover)
ISBN 978-981-126-431-3 (ebook for institutions)
ISBN 978-981-126-432-0 (ebook for individuals)

For any available supplementary material, please visit
https://www.worldscientific.com/worldscibooks/10.1142/13084#t=suppl

Desk Editor: Joseph Ang

Typeset by Stallion Press
Email: enquiries@stallionpress.com

To Dvora, my beloved wife,
companion and friend throughout my life

Contents

Introduction

Sooner or later, everyone asks the big questions.

- Where did the universe come from?
- Where did our planet come from?
- Where did life come from?
- Where did human beings come from?

As is well known, Scripture (Genesis) has answers to these questions. And we will be discussing these answers on these pages. And what does science say? We shall see that science also has something to say about all these questions.

In recent years, many branches of science have been revolutionized. Completely new disciplines now occupy a central place in modern scientific thought. Does this new scientific knowledge have any implications for understanding the Bible? The goal of this book is to demonstrate that the implications are profound. We will show that accepting the truth of the Bible does not conflict with rational thinking. Quite the contrary. This book will demonstrate how modern scientific topics, including quantum theory, chaos theory, the butterfly effect, string theory, the multidimensional universe, and supernova explosions, all play a role in understanding biblical passages. Moreover, these scientific topics will be explained clearly and simply, in a way that can be readily understood by an intelligent reader even without any scientific background.

I am an Orthodox Jew. It is certainly not the case that all Orthodox Jews share the same opinions regarding Scripture, but we all attribute a divine origin to Scripture. Therefore, let me say frankly to anyone who thinks that the Book of Genesis is simply a series of mythological tales, assembled over the centuries in various ancient societies, this book is not for you. However, for those who take Scripture seriously, this book will demonstrate how modern science can shed new light on the words of Scripture.

Chapter 1

The Relationship Between Science and Scripture

In this chapter, we shall consider several approaches that have been proposed over the years regarding the relationship between science and Scripture.

Does Science Contradict Scripture?

The idea that there is a confrontation between science and Scripture became popular in the nineteenth century, when many books appeared claiming that there exists an ongoing war between science and Scripture. Most prominent among these writers were John William Draper (*History of the Conflict between Religion and Science*) and Andrew Dickson White (*The Warfare of Science*). And, of course, there was the famous confrontation regarding evolution in the 1860 meeting of the British Association of Science, pitting Bishop Samuel Wilberforce against Thomas Huxley ("Darwin's bulldog").

The intensity of the Science vs. Religion wars has lessened recently, but has not disappeared entirely. In 2006, Richard Dawkins, of the University of Oxford, who holds the dubious title of being the world's most famous atheist, published *The God Delusion*, which has sold millions of copies. Dawkins has repeatedly claimed that science has shown the irrationality of Scripture, famously stating[1]: "Darwin kicked God out of biology." However, Charles Darwin did not agree with Dawkins's provocative statement. In fact, he ended his famous 1859 book, *On the Origin of Species*, with the following stirring words:

> *"There is a grandeur of this view of life, having been originally breathed by the Creator into a few forms or into one, and from so simple a beginning, endless forms most beautiful and most wonderful have been and are being evolved."*

Note the capital "C" in the word "Creator," meaning, of course, God. In other words, Darwin viewed his theory of evolution as the *mechanism* used by God to produce the animal and plant kingdoms. This is often referred to as "theistic evolution."

Does Science Confirm Scripture?

In complete contrast to the idea that science has disproved Scripture, some writers have taken the exact opposite view, claiming that science has proved the validity of Scripture! Most prominent among these writers was the British theologian William Paley. In 1802, Paley wrote the following[2]:

> *"If one were to walk in the forest and find a rock, one could imagine that the rock has always been there and had not been made by anyone. However, if one were to find a watch in the forest, no one would suggest that the watch had always been there and had not been made by a person. The precision with which the cogs, springs, and gears of the watch have been fashioned, and the intricacy with which these have been assembled to serve a particular purpose all demonstrate that the watch could not have been formed by natural processes. Rather, its complexity and specific design prove that the watch must have been made by a watchmaker."*

---

[1] Quoted in the newsmagazine *The Economist* 5 September 2010, Science and Religion section.

[2] W. Paley, 1802, *Natural Theology*, part of the seven-volume Works, vol. 5, pp. 1–2.

"If one now considers the natural world, with its vast panorama of animals and plants, each consisting of many complex patterns of tissues and organs that function together in intricate ways to permit each animal and plant to live and bear young, one sees far more complexity than is found in any watch. Therefore, if the complex design of a watch requires a watchmaker, how much more so must the complex design of the natural world require a 'Maker,' who must be God."

Paley's "scientific proof" for the existence of God and, thus, the validity of Scripture, is called the "argument from design" or often, to honor Paley, "the watchmaker argument." (The word "argument" is an old English word for "proof.")

Although many persons are still convinced by Paley's argument from design, the proof is fallacious. In fact, many extremely complex entities form all by themselves, without the intervention of a "maker." Consider snowflakes, which are crystals of snow in the form of beautifully intricate structures, no two of which are alike, each having a perfect six-fold pattern of fractal symmetry. Nevertheless, as we all know, the delicate beauty of snowflakes is *not* designed. Snowflakes form spontaneously under certain weather conditions.

There are many other examples of extremely complex objects that form all by themselves under certain conditions. Any chemist can list unbelievably complex molecules that form spontaneously whenever the required raw materials are present at an appropriate temperature. Further examples of complex structures that form spontaneously include the vertex structure of type II superconductors, intricate crystals, fractals, self-similar repeating patterns, Mandelbrot sets, and many more. Therefore, it is clear that the existence of complex objects does not provide any evidence that the universe was designed.

Consider now the following counter-example to the argument from design. Imagine walking in the woods and finding the letters ABC carved on a tree trunk. You would immediately conclude, and rightly so, that someone had carved these letters into the wood. There is nothing complex about the letters ABC, just a few lines arranged in a very simple pattern. Nevertheless, it is clear that some conscious person had carved these letters.

Our *experience* tells us that letters *never* form spontaneously via the processes of nature, but are *invariably* written by a person. Similarly, *our experience* tells us that watches never form spontaneously, but are always produced by a watchmaker. It is not complexity, but experience that determines which objects form spontaneously and which objects require intention for their production.

Let us now consider the universe. Was it fashioned by intent or did it come into existence spontaneously? Here, we have no experience to guide us because *only one* universe exists. Therefore, we are unable to conclude *anything* about the origin of the universe, regardless of its extreme complexity.

Science and Scripture Are Very Different Subjects

A very different approach has appeared in recent years, claiming that there is no conflict between science and Scripture because these are two completely separate realms and, therefore, neither can negate the other.

In his 2011 book, *The Great Partnership*, Lord Rabbi Jonathan Sacks wrote, in his usual clear and graceful style:

> *"Science and Scripture are as unlike as poetry and prose, or song and speech, or a portrait of a person and an MRI scan. Science takes things apart to see how they work. Scripture puts things together to see what they mean. Science is about explanation, Scripture is about interpretation. Science analyses, Scripture integrates. Science breaks things down to their component parts. Scripture binds people together in relationships of trust. Science tells us what is, Scripture tells us what ought to be. Science describes, Scripture inspires, beckons, calls. Science sees the underlying order of the physical world. Scripture hears the music beneath the noise. Science is the conquest of ignorance. Scripture is the redemption of solitude."*

Lord Sacks was preceded in this approach by Stephen Jay Gould, of Harvard University. In a widely quoted essay, entitled

"Nonoverlapping Magisteria," Gould[3] states that science and Scripture deal with totally different topics and, therefore, one shouldn't expect to find scientific truths in Scripture.

Professor Gould himself was preceded in this approach by Galileo, who penned the famous aphorism[4]: "Scripture teaches us how to go to heaven, whereas science teaches us how the heavens go." This sentence appeared in his 1615 letter to the Grand Duchess Christina of Tuscany. (These words sound as snappy in Galileo's Italian, as they do in English.) Galileo was hauled before the Court of the Inquisition, who held that it certainly *was* within the authority of the Church to use Scripture to "teach us how the heavens go." (The Court sentenced Galileo to life imprisonment, but Pope Urban VIII, who was Galileo's friend, immediately commuted the sentence to house arrest. Since Galileo lived in a fancy villa, this sentence was not very harsh.)

One of the most outspoken contemporary expositors of the Galileo–Gould–Sacks approach was the philosopher-scientist Yeshayahu Leibowitz, who repeatedly asserted that the Scripture has nothing at all to say about the physical world ("Genesis is not a science book"). In particular, according to Leibowitz, the creation narrative in the first chapter of Genesis is *not* an account of the origin and development of the universe. Rather, it is discussing the relationship between man and God.

Professor Leibowitz said the same thing about the biblical account of history ("Scripture is not a history book"). If Scripture is not saying anything about the patriarchs, the Flood, the Exodus from Egypt, or the remarkable capture of Canaan by the Israelites, then it follows that there are no historical questions that need answering.

The problem with the Sacks–Galileo–Gould–Leibowitz approach is obvious. Scripture *does* make statements about the physical universe — quite a lot of them, in fact — and, therefore, it is

---

[3] S. J. Gould, March 1977, *Natural History*, pp. 16–22.
[4] P. Redondi, 1983, *Galileo: Heretic* (transl. R. Rosenthal, Princeton University Press: Princeton).

perfectly legitimate to inquire how these statements bear up under scientific scrutiny.

A more modest version of this approach (that Scripture is not discussing the physical universe) is to restrict it to the creation narrative. Many traditional Jewish biblical commentators have stated that the Creation is shrouded in divine mystery and lies beyond human understanding. For example, Moses Nachmanides[5] (twelfth century) calls the Genesis creation narrative "a deep mystery, which cannot be understood by reading the verses."

But this does not solve the problem because Scripture-vs.-Science contradictions are not confined to the first chapter of Genesis. The fifth chapter records thousand-year life spans. The seventh chapter describes a universal Flood. The 35th chapter states that Isaac lived to the age of 180. And we read in Exodus that within a couple of hundred years, Jacob's family exploded into a nation nearly 10,000 times as numerous. Thus, Scripture-and-science problems appear everywhere. Excluding the first chapter of Genesis from consideration does not solve the problem.

Scripture Is Eternal but Science Is Transitory

Another approach to the relationship between Scripture and science is to claim that whereas the divine words of Scripture are eternal, science is always in a state of flux. Indeed, change is thought to be the most conspicuous feature of science, and it is claimed that all scientific theories will eventually become discarded, to be replaced by new paradigms.

The facts are quite otherwise. Every competent scientist can distinguish between speculative theories and those that are supported by a vast array of scientific evidence. The former are short-lived, whereas the latter have an excellent record for longevity. For example, the theory of relativity and the quantum theory have enjoyed unqualified success in explaining hundreds of physical phenomena.

---

[5] Nachmanides, *Commentary on Genesis*, 1:3.

Well-established theories become refined, but they are never simply discarded as being wrong.

The excellent track record of well-established scientific theories was emphasized by Nobel laureate Steven Weinberg[6]: "One can imagine experiments that *refute* accepted scientific theories. *There are no such examples whatsoever in physics in the past hundred years*" (emphasis in original). *None whatsoever!*

What about the geocentric theory of the solar system? Wasn't that scientific theory universally believed for 1500 years, until it was finally shown by Copernicus and Galileo to be wrong and then replaced by the very different heliocentric theory?

The answer is: "No." The geocentric theory was not a scientific theory at all. The theory was universally accepted for over a millennium on religious grounds alone. The *beliefs* of the Church demanded that man's place *must be* at the center of the universe. *Religious beliefs* required that planetary orbits *must be* circular because the circle is the ideal geometric figure and God's heavens must behave in the ideal manner. Even when circular orbits failed to explain the planetary data, circles were not abandoned. Additional circles, called *epicycles*, were simply added to the geocentric theory. Eventually, *eighty* epicycles were postulated, but the details of planetary motion *still* could not be explained. Finally, when the telescope was invented in the early seventeenth century, accurate astronomical data became available, confirming the *first scientific theory* of planetary motion — the heliocentric solar system with elliptical planetary orbits rotating around the sun, located at one focus of the ellipse — a scientific theory that has survived unscathed to this very day.

What about Newton's theory of mechanics? Wasn't Newtonian mechanics overturned by relativity theory in 1905 and overturned again by quantum theory in 1926?

Not at all! Newtonian mechanics was *generalized* by these theories, and shown to be *correct* for low velocities (even thousand kilometers per second is *slow* in this context!) and for large masses

---

[6]S. Weinberg, 1993, *Dreams of a Final Theory* (Vintage: London), p. 102.

(even a speck of dust weighing a trillionth of a gram is a *large mass* in this context!). Newton's theory is so accurate within its wide regime of validity that every university student of physics is required to learn Newtonian mechanics.

Finally, we have the assertion of some philosophers that one cannot have confidence in science because scientific theories are empirical and based on induction. These theories are derived from a finite number of data points. However, there exist an *infinite* number of theories that can explain any *finite* set of data. (In technical language, one can draw an infinite number of continuous curves through any finite number of points.) Therefore, there is no reason to think that today's scientists were lucky enough to have chosen the correct theory from the infinite number of possibilities.

Research scientists are amused by the assertion that for each set of measurements, there are a large number of theoretical explanations just waiting for the scientist to choose the theory that strikes his or her fancy. This idea is utterly removed from reality. In truth, scientists spend most of their time struggling to formulate *any theory* that might explain the observed data, or at least part of the data.

Scientists will never achieve the "final truth," but there are good reasons for thinking that our understanding of the physical world is becoming progressively more and more accurate.

There Are No Contradictions

A new "solution" has recently been proposed for reconciling contradictions between science and Scripture. *There are no contradictions!* It is claimed that a true understanding of science enables one to explain *all* apparent discrepancies in the Genesis account of creation. An obvious discrepancy relates to the time scale.[7] How does one reconcile the "Six Days of Creation" of Scripture with the multibillion-year-old universe of the scientist?

According to Einstein's theory, time is relative. The rate at which time passes varies from place to place in the universe according to the local strength of gravity — an effect called "time dilation."

---

[7]G. L. Schroeder, 1991, *Jewish Action*, pp. 44–48.

To explain the Six Days of Creation of Scripture, one assumes that the "Genesis clock" was located at a place where large gravitational forces were present. Therefore, time passed much more slowly, with the "Genesis clock" advancing only six days while 14 billion years were passing elsewhere in the universe.

This explanation fails because time dilation is an extremely small effect. For example, a year measured on the very massive Sun is only *one minute* shorter than a year measured on Earth, a change of only two parts per million. Such a small effect cannot possibly compress 14 billion years into a mere six days. There is no place in the universe where extremely large gravitational forces exist now or where they have ever existed in the past. Indeed, this proposed explanation for the Genesis time scale was characterized by a distinguished physicist as "a fundamental misunderstanding of basic physics, immediately recognized as fallacious by professional physicists."[8]

Recently, an attempt was made to reconcile Scripture with the scientific time scale by invoking the "cosmological red shift." (This is the change in the color of spectral lines of distant galaxies that are receding from us.) This shift in the color of light is the analogue of the shift in the pitch of sound, an effect proposed by Christian Doppler back in 1842. There is a well-known example of the Doppler shift for sound. As a train approaches the station, the pitch of the sound made by the train increases to higher notes, and as the train recedes from the station, the pitch of the sound made by the train decreases to lower notes. A change in pitch for sound is equivalent to a change in color for light, and is therefore called the "red shift." However, this shift in the color of spectral lines has *nothing to do with a change in the rate at which time passes.*

It Just Looks Old

A quite imaginative "solution" to reconcile Scripture with the scientific time scale is to assert that the universe is, in fact, young, just as Scripture states, but it *appears old* because God created the universe

---

[8]B. Simon, Spring 1992, *Jewish Action*, p. 10.

to look old. For example, if Adam were created by God as a 20-year-old man, then the day after his creation, a physician who examined him would declare that Adam had already lived for twenty years. It is similarly asserted that dinosaur bones and other fossils were created *recently* and then placed in the ground by God. These fossils lead scientists to conclude, erroneously, that the Earth is many millions of years old.

Although it is impossible to prove that this approach is wrong, a satisfactory reconciliation between science and Scripture should *not* invoke miracles. Moreover, one might wonder just what divine purpose was served by placing so many ancient-looking dinosaur bones in the ground.

Approach of Maimonides

Having criticized all other approaches, it is time to reveal my preferred solution. In one of the most important essays ever written in the area of Science and Scripture (*Guide for the Perplexed* 2:25), Moses Maimonides (twelfth century) explains how to interpret the narrative portions of Scripture. Maimonides states that one should always first attempt to understand the words of Scripture literally. However, if the literal meaning contradicts well-established knowledge, then one should interpret the words of Scripture *metaphorically*, because, as Maimonides wrote, *"the paths of interpretation are not closed to us."*

We may apply Maimonides's principle to the time scale. If one interprets the Six Days of Creation literally, as six periods of 24 hours each, then a wealth of well-established scientific knowledge is completely contradicted. Therefore, following Maimonides, one understands the *"days of creation"* metaphorically, referring to eras or phases in the development of the universe, without any indication of the length of each phase.

In the twelfth century, the accepted view was that the universe was eternal and no act of creation had ever occurred — which, of course, completely contradicts the first chapter of Genesis. Maimonides states that he rejects the eternity of the universe, but *not because it disagrees with the words of Genesis.* Rather, he rejects this view because it was not convincingly proved. However, states Maimonides, if this view were to be proved, then, from the traditional Jewish point of view, there would be *no problem* in accepting the idea of an eternal universe, *in spite of the fact* that a denial of creation flatly contradicts Genesis. One would simply understand the entire first chapter of Genesis figuratively, as an allegory.

Maimonides is not the only major traditional Jewish commentator to assert that the biblical word *day* need not always be understood literally. The Book of Hosea discusses (6:2) *the two days* and *the third day*. The Jewish commentator Rashi comments on this verse that each of these three "*days*" mentioned in the Book of Hosea refers to a different *period* in Jewish history.

Chapter 2

Scripture, Science, and Creation I: The Origin of the Universe

The creation narrative in the first chapter of Genesis contains many apparent contradictions between science and Scripture. There are three main topics, each of which deserves a separate chapter:

- **Cosmology**: What is the origin of the universe? Was the universe indeed created from nothing, as stated in Scripture?
- **Evolution**: Is the animal kingdom the result of gradual evolutionary processes, lasting millions of years, as scientists claim? Is this idea compatible with Scripture?
- **Appearance of Man**: Did human beings arise through gradual evolutionary processes from an ancient monkey-like creature, as scientists claim? Is this idea compatible with the statement in Scripture that God *created* man?

This chapter deals with cosmology, the first of these important topics. We begin by writing the first five verses in Genesis — Day One of the creation narrative.

Verse 1: "*In the beginning, God created the heaven and the earth.*"
Verse 2: "*The earth was chaotic, and darkness covered the abyss. The spirit of God hovered over the water.*"
Verse 3: "*God said, 'Let there be light,' and there was light.*"

Verse 4: "*God saw that the light was good. God separated the light from the darkness.*"

Verse 5: "*God called the light 'day' and He called the darkness 'night.' There was evening and there was morning, one day.*"

Some comments must be made before we relate these words to science.

- The expression *heaven and earth* in Verse 1 is a common Scriptural expression for the entire world. Also in modern English, these words often have this meaning: "I will move heaven and earth to accomplish this task."

- The Hebrew words *tohu va-vohu* in Verse 2 have been translated as "chaotic." This corresponds to the Hebrew translation of the English word "chaotic" that is found in English–Hebrew dictionaries. The importance of this translation will soon become clear.

- In those verses that attribute an act to God (such as, *God separated the light from the darkness*), we are interested only in the act itself (light was separated from darkness), without any reference to the divine being to whom the act is attributed. Thus, the question before us is whether scientists confirm that such an act actually did occur. It should be mentioned that the scientists whom we shall quote are all secular and have no interest in confirming the existence of God.

- Not every word or phrase in these five verses has physical content that can be compared to science. For example, we have nothing to say about words such as, *God called the light "day" and He called the darkness "night"* or *The spirit of God hovered over the water* or *God saw that the light was good.* However, we shall have much to say about those words and phrases that do have physical content.

The Origin of the Universe

Where did the universe come from? A person of faith would probably answer that the universe was created, as stated in the first verse of Genesis. Such an answer was long considered a scientific impossibility, because it contradicted the law of the conservation of matter and energy. According to this law of science, which was established in the middle of the nineteenth century, both matter and energy can be changed from one form into another form, but *something* cannot come out of *nothing*. Therefore, scientists viewed the universe as eternal, thus neatly avoiding questions regarding its origin. The Genesis assertion that the universe was created, presumably from nothing, became an area of conflict between science and Scripture. That is how matters stood for many years.

This situation has now changed. The twentieth century witnessed an explosion of scientific knowledge, especially in cosmology, the discipline that deals with the origin of the universe. For the first time, advances in cosmology during the past few decades have permitted scientists to construct a coherent history of the origin of the universe. Today, an overwhelming body of scientific evidence supports the Big Bang theory of cosmology.

The following scientific evidence supports the Big Bang theory: (1) the discovery of the remnant of the initial ball of light (or popularly, the big bang), called the cosmic microwave background radiation because today, this light lies primarily in the microwave portion of the spectrum, (2) the measured hydrogen-to-helium ratio throughout the universe, (3) the observed expansion of the galaxies, and (4) the perfect agreement between the predicted and measured spectrum of the cosmic microwave background radiation, as found by two space satellites: COBE (1989) and WMAP (2001). The term "spectrum" refers to the ratio of the intensities of the different colors of the measured light. For example, how much blue light is there relative to the amount of red light?

Only the Big Bang theory accounts for all these observations and, therefore, the academic scientific consensus accepts this theory. According to the prestigious journal for the layman *Scientific American*[1]: "The Big Bang theory works better than ever."

The central theme of the Big Bang theory is that *the universe began through an act of creation*. It is instructive to quote some of the world's leading authorities on the subject of creation.

---

[1] G. Musser, February 2004, *Scientific American*, p. 30.

Nobel laureate Paul Dirac, a leading physicist of the twentieth century, writes[2]:

"It seems certain that there was a definite time of creation."

Cosmologist Alan Guth, of the Massachusetts Institute of Technology, writes[3]:

"The instant of creation remains unexplained."

Joseph Silk, an important cosmologist from the University of California, writes[4]:

"The Big Bang is the modern scientific version of creation."

Physicist Brian Greene, of Columbia University, writes[5]:

"The currently accepted scientific theory of creation is often referred to as the standard model of cosmology."

Finally, cosmologist Stephen Hawking, of the University of Cambridge, writes[6]:

"The creation lies outside the scope of the known laws of physics."

The term **creation** has left the private preserve of the biblical scholar and has entered the lexicon of science. Today, it is not possible to carry on a meaningful discussion of cosmology without the creation of the universe playing a central role. We note that to the scientists, the term "creation" has no religious connotation. It simply

---

[2] P. Dirac, 1992, *Commentarii*, vol. 2, no. II, p. 15.
[3] A. Guth, May 1984, *Scientific American*, p. 102.
[4] J. Silk, 1989, *The Big Bang* (W. H. Freeman: New York), p. xi.
[5] B. Greene, 1999, *The Elegant Universe* (Johnathan Cape: London), p. 15.
[6] S. Hawking, 1973, *The Large Scale Structure of Space-Time* (Cambridge University Press: Cambridge), p. 364.

means that the universe had a beginning. *But this is exactly what is written in the first verse of Genesis.*

The Light

When cosmologists speak of the "creation," to which event are they referring? Exactly what was created to begin the universe? Scientists have discovered that the universe began with the sudden appearance of an enormous *ball of light*, which scientists call the "primeval light-ball," but which was dubbed the "Big Bang" by the British astrophysicist Fred Hoyle and hence the name of the theory. The remnant of the initial light-ball was detected in 1965 by two American scientists, Arno Penzias and Robert Wilson, who were awarded the 1978 Nobel Prize for their important discovery.

The discovery of the primeval light answers another long-standing puzzle regarding the Genesis account of creation. On the First Day of Creation, Scripture states (Verse 3): *and there was light.* But at that time, there existed neither stars, nor sun, nor any other known source of light. Therefore, how can one understand the Genesis "light"?

For thousands of years, this question remained unanswered. Scientists have now discovered that *there was light at the very beginning of time* — the primeval light-ball whose appearance heralded the origin of the universe. This light did not appear *within* the existing universe. Rather, the creation of light *was* the creation of the universe. *This is exactly what is written in Genesis.* That is, Genesis does not record *two* separate creations on the First Day — the creation of the universe *and* the creation of light — but *only one.*

Separation of the Light From the Darkness

The words in Verse 4, *God separated the light from the darkness,* seem problematical. "Darkness" is not a physical entity that could have been "separated from the light." The word darkness simply denotes the absence of light. Therefore, what meaning can one give to these strange words?

To answer this question, one must explain some more details of the Big Bang theory. We have already noted that the universe began with the sudden appearance of an enormous ball of light, the big bang. Immediately after this light-ball appeared, it was the only object in the universe. However, the present universe is filled with matter, ranging from stars and planets to oceans, trees, and animals. Where did all this matter come from if nothing originally existed but the enormous ball of light?

The answer is given by Einstein's famous equation $E = Mc^2$. The letter E denotes energy. The light-ball was an enormous source of energy. The letter M denotes mass or matter and the letter c denotes the speed of light. According to Einstein's equation, it is possible under appropriate circumstances to transform matter into energy (which is the source of nuclear energy and of nuclear bombs). However, the equation works both ways. It is also possible under appropriate circumstances to transform energy into matter. All the matter that exists in the universe today derives from the transformation of most of the enormous energy of the light-ball into matter according to Einstein's equation.

The familiar form of matter is atoms, or groups of atoms called molecules. All the matter with which we are familiar, including planets, rocks, water, and air, is composed of atoms and molecules. However, when matter was first formed from the initial light-ball through Einstein's equation, the matter did not exist in the form of atoms. The enormous temperature of the light-ball would have instantly disintegrated any atom. Therefore, the initial matter existed in a different form, called "plasma." The important difference between these two forms of matter, atoms and plasma, is the following. An atom is electrically neutral, consisting of an equal number of positive charged particles (protons) and negatively charged particles (electrons). In contrast to this, plasma consists of charged particles, either electrons by themselves or protons by themselves.

An important property of charged particles is that they "trap" light and prevent its free passage. Therefore, as soon as plasma was formed, the universe became dark. The universe consisted of the remnant of the light-ball interspersed with plasma. Even though the light was very intense, it was "trapped" by the plasma and could not escape to be "seen." Therefore, the universe became dark.

The initially extremely hot universe cooled very rapidly and soon became cool enough for the negatively charged electrons to coalesce with the positively charged protons to form electrically neutral atoms. When that happened, the light was no longer trapped by the dark plasma, and suddenly burst forth, filling the universe with light. This is the situation that remains to this very day.

With this background, the words in Verse 4 relating to *the separation of the light from the darkness* can be understood as referring to the separation of the light from the previously dark plasma. This event was of vital importance in the development of the universe into a suitable place for human beings. If there was only plasma in the universe, there could be no planets, no water, no air, and of course, no living creatures. All of these items are composed entirely of atoms and molecules.

Chaos = "Tohu Va-Vohu"

An important development in the Big Bang theory falls under the general heading of the "inflationary universe." It lies beyond the scope of this book to explain the nature of these new findings. However, we note that according to an article of this topic[7]: "The universe began in a random chaotic state." Professor Andrei Linde, of the Lebedev Physical Institute in Moscow, has proposed the "chaotic inflation scenario" to describe the beginnings of the universe. In short, the role of chaos in the development of the early universe has become a major topic in cosmological research.

The relevance of this to our discussion is clear. Genesis asserts that the universe began in a state of chaos (in Hebrew: *tohu va-vohu*).

In summary, we see that every word and every phrase regarding physical aspects of the universe that appears in the Scripture description of the creation (the five verses of Genesis) has a counterpart in the modern Big Bang theory of cosmology.

---

[7] A. Guth, May 1984, *Scientific American*, p. 102.

Water Above the Heaven

The next three verses (Verses 5–7, Second Day of Creation) deal with the formation of the heaven — what a scientist would call outer space. Verse 7 describes the function of the heaven: *God separated the water below the heaven from the water above the heaven.* The phrase *water below the heaven* is readily understood to mean lakes, oceans, rivers, and regular water, but what can be meant by the phrase *water above the heaven*? Has anyone ever heard of large bodies of water floating about in outer space?

Since the space program in the 1970s, we are no longer restricted to peering into outer space from the surface of our planet and speculating about the solar system. Today, scientists launch satellites into space, make measurements, and bring the data back to earth. As a result, for the first time in history, we understand the detailed composition of the solar system. One of the most exciting discoveries was the existence of enormous quantities of frozen water at the edge of the solar system.

The principle source of water in outer space is the comets. Scientists refer to comets as "dirty snowballs," because a comet is essentially an enormous block of ice interspersed with interplanetary debris. The numerical data relating to comets staggers the imagination. A small comet contains a *billion tons of ice.* A large comet contains a *thousand times as much.* There is a vast reservoir of comets at the outer edge of the solar system called the Oort reservoir, named after the Dutch astronomer Johannes Oort who discovered it. The Oort reservoir contains *hundreds of billions of comets*, with a typical comet containing *hundreds of billions of tons of ice.* If it were possible to bring all this ice to our planet and melt it, the resulting water would fill the ocean basins one thousand times! *The water above the heaven* really exists, *just as is written in Genesis.*

Remainder of the Genesis Creation Narrative

The agreement described above between science and Scripture relating to the First Two Days of Creation continues throughout the remaining Days of Creation. Recent scientific discoveries show that every word and phrase in the first chapter of Genesis that has physical content agrees with modern science, including cosmology, astronomy, geology, biology, and archaeology. Scripture and science are indeed in harmony.

Chapter 3

Scripture, Science, and Creation II: The Origin of the Animal Kingdom

One of the important news stories in the area of science and religion was the 1999 decision of the Kansas Board of Education (later overturned) to remove the subject of evolution from its high school education standards. Accordingly, high school students in Kansas would no longer be tested on the subject of evolution, which of course guaranteed that they would no longer be taught the subject. An election in Kansas resulted in a Board of Education with a majority of creationists. The new board opposed teaching evolution as fact, and they insisted that evolution should be taught as "merely a theory" — meaning, an unproved speculation. The creationist view of the origin of the animal kingdom — the separate divine creation of every single species — was to be taught as an equally acceptable explanation.

One of the main planks in the belief system of creationists is that biological evolution never occurred, and that it is a secular idea proposed by atheists to turn people away from belief in the divinity of Scripture. However, disbelief in evolution is not restricted to creationists. Poll after poll shows that about half the adult population in the United States does not believe in the evolutionary origins of the animal kingdom.

We shall here examine *why* creationists are so opposed to the scientific concept of evolution. There are many other phenomena for which the scientific explanation does not correspond to the literal words of Scripture, but these other subjects do not provoke creationists into demanding changes in the school curriculum.

Consider the rainbow. Science classes teach the Newtonian theory which states that the rainbow is a natural phenomenon, caused by white sunlight being separated into its various colors by drops of rain which act as prisms. But according to Scripture (Genesis 9:13), the rainbow is of divine origin. Yet, one never hears of creationists insisting that the Newtonian theory of the rainbow be taught as "merely a theory," and that equal time be devoted in the classroom to the biblical account of the rainbow. What is so offensive in their eyes about evolution that generates such spirited opposition?

Heliocentric Solar System

There is another scientific issue that, in its time, generated even more controversy. This is the question of whether the sun revolves around the earth (geocentric theory) or whether the earth revolves around the sun (heliocentric theory). The seventeenth-century Italian astronomer Galileo was put on trial by the Inquisition Court of the Catholic Church for championing the heliocentric theory. Under threat of death, Galileo was forced to publicly recant.

Opposition to the heliocentric theory was not limited to the Catholic Church. *Encyclopedia Judaica* states[1]: "The Jewish writings on astronomy of the eighteenth century and the rabbinical literature of the nineteenth century are basically derived from the geocentric theory. In his book, *Ma'aseh Tuvia* (Venice, 1708), Tuvia Cohn presents the geocentric theory in its classic form. The heliocentric view is also analyzed, but is rejected on religious grounds."

It is quite unexpected that there should be a religious controversy regarding which heavenly bodies are stationary and

---

[1] A. Beer, 1972, *Encyclopaedia Judaica* (Keter Publishing House, Jerusalem), s. v. "Astronomy", vol. 3, p. 805.

which move. The Genesis account of creation does not favor the geocentric system. Just where in Scripture is it written that the sun revolves around a stationary earth? This matter is touched upon in the Book of Joshua. During the battle in the Valley of Ayalon, Joshua asked God to command the sun and moon "to stand still" so that the battle could be completed victoriously while there was still daylight (Joshua 10:12–13), and these heavenly bodies did so. Since Joshua asked God to command the sun and moon to stand still, it follows — so goes the argument — that under ordinary circumstances, the heavenly bodies are *not* stationary, but revolve about the earth. On the basis of this flimsy exegesis, the Church condemned Galileo to life imprisonment and ordered his books to be burned! Surely, there must be something far deeper here.

The Age of the Universe

The third phenomenon that arouses great passion in the creationist camp is the age of the universe. Creationists are completely opposed to the 14-billion-year-old universe of the cosmologists, and insist that the Six Days of Creation of Genesis are to be understood literally as six 24-hour periods of time. It is easy to show that the creationist insistence on this point is not based on the view that words in Scripture must be understood literally.

Consider, for example, the "light" that is mentioned on the First Day of Creation. On the First Day, there were no stars, no sun, no people, and no known source of light. Creationists explain that the Genesis "light" was not *physical* light at all, but rather *spiritual* light of some sort. No creationist seems troubled by the fact that the literal meaning of Scripture has thereby been set aside in favor of a figurative interpretation. In fact, creationists interpret many biblical verses as meaning something very different from the literal text.

What do the creationists find so unacceptable about an ancient age for the universe? Why do they reject interpreting the Six Days of Creation *figuratively*, as spiritual days of undefined length, just as they explain the Genesis light *figuratively*, as spiritual light?

Man's Spirituality

It is a cardinal principle of Scripture (Genesis 1:28–29) that mankind was the ultimate purpose of God's creation — that human beings are spiritual creatures who hold a position of central importance in the universe. What are the implications of this principle for the believing person?

If mankind is of central importance *spiritually*, then our spirituality would seem to imply that mankind cannot have developed physically from a simple bacterium. Such an approach would also explain the creationist rejection of the theory of evolution, which states that mankind shares physical origins with the cockroach and the crocodile.

Similarly, such an approach would also imply that the planet we inhabit must lie at the very *center* of the universe, with the other heavenly bodies whirling around it. Placing the earth among the other planets, as just another astronomical body that revolves around the central sun, appears to be an affront to the spirituality of man.

Historically, it is clear that it was not for scientific reasons that the geocentric theory of the solar system went unquestioned for 1400 years. The appeal of this theory was not based on scientific considerations. As one book on the history of science puts it[2]:

> *"The universal popularity of the geocentric theory of the universe was due to the important place it gave man in the general scheme of things. The geocentric theory added immensely to man's already developed sense of his own importance. As astronomical observations increased in accuracy, more and more complicated assumptions were simply added to the geocentric hypothesis to explain the astronomical data."*

Further support for the view that the geocentric theory of astronomy was not based on scientific considerations can be found in the trajectories assumed for the heavenly bodies. It was taken for granted that the sun, moon, planets, and stars all revolved about the central earth in *circular orbits*. Even Copernicus, who proposed the

---

[2] J. Clarke, 1954, *Man and the Universe* (Simon and Schuster: New York), p. 27.

revolutionary idea of a heliocentric solar system, still assumed circular orbits for the planets. What was so attractive about the assumption of circular orbits? The correct shape of the planetary orbits, an ellipse, was a well-known geometric figure, one of the conic sections that had been studied in detail by the ancient Greeks. Why did no one consider the possibility that the heavenly bodies might move in elliptical orbits?

The explanation is to be found in the prevailing view of the universe. Since it was assumed that God Himself controlled the motion of the heavenly bodies, their orbits must be "perfect." The ideal geometric figure has long been considered to be the circle. It followed, therefore, that the divinely controlled orbits of the heavenly bodies *must* be circles. Even when it became obvious that circular planetary motion was inconsistent with astronomical observations, the circle was never abandoned. It was still taken for granted that the planets moved in circles, but it was now assumed that the center of the circle (called an "epicycle") moved on a different circle. And when even this more complicated theory proved inadequate to explain the ever more accurate data, it was assumed that the circular epicycle must revolve around yet another circular epicycle, which itself moved in a circle. Planetary trajectories could *never* be anything but various combinations of circles, however complicated, because theological considerations restricted medieval astronomers to ideal geometric figures. As a book on the history of science explains[3]:

> *"By the year 1500, over 80 epicycles were required to account for the motions of the five known planets, the sun, and the moon. That astronomers of that day were able to devise so intricate a picture to account for the observed facts is a tribute to their mathematical ability and ingenuity, but not to their scientific judgment."*

This astronomical theory was but one aspect of a comprehensive theological approach to the physical world, which was assumed to be a window to God. However, by the 1600s, the scientific community

---

[3] J. Clarke, 1954, *Man and the Universe* (Simon and Schuster: New York), p. 28.

finally came to realize that the geocentric theory simply could not explain the detailed observations of the planets, no matter how many new epicycles were added. The heliocentric theory gradually carried the day, and mankind was relegated to a "minor" planet, far from the center of the universe.

Just as the creationists became reconciled to this state of affairs, another blow fell. In 1859, Charles Darwin published his famous book, *The Origin of Species*, introducing the theory of evolution by natural selection. This theory asserts that the vast panorama of animals and plants, *including man*, all developed from simpler forms. Scientists had already shown that human beings did not occupy a central planet in the *physical* world. And now scientists were claiming that man is not a special species in the *biological* world. Creationists eventually did come to terms with the heliocentric solar system. Indeed, it would be absurd to deny it today, in the age of space exploration. But, regarding evolution, they continue to assert the biological uniqueness of human beings.

As the preceding discussion makes clear, it is *not* Scripture that underlies the creationist opposition to evolution, but a matter of their *weltanschauung*.

According to the creationist perception of human spirituality, mankind must have had a different physical origin from that of the "lower" animals. This perception sees human dignity as being degraded by the implication that "man's ancestor was a monkey." The creationist concept places each species in its divinely assigned niche, and thus maintains the unique spiritual position of mankind.

Creationists are also completely opposed to the idea of a very ancient universe. Human civilization, the only subject of interest in Scripture, began only a few thousand years ago. If the universe has existed for more than 10 billion years, then the extremely long period before mankind accounts for 99.9999% of the history of the universe. This poses a serious problem for the creationist. If man is really so important, why was his creation delayed for more than 10 billion years?

This question shows how the existence of an ancient universe *appears* to cast doubt on the divinity and spirituality of human beings. Once again, it is *not* the text of Genesis, but the creationist perception and understanding of the text that prevents acceptance of the modern cosmological scenario. Their outlook seems to imply that the *physical* absence of mankind over such a large part of the history of the universe indicates that man lacks *spiritual* worth. It implies that God permitted the universe to exist for billions of years without any purpose whatsoever. Since this is impossible in their view, it follows that the universe *cannot* be as ancient as the cosmologists maintain.

The conflict over the heliocentric theory differs in an important aspect from the conflict over evolution and an ancient universe. In the twenty-first century, knowledge of astronomy and achievements in space flight make it impossible to deny the heliocentric theory. However, the origin of the universe and biological evolution deal with events that occurred in the far distant past, when no human beings were present. Therefore, so the creationists claim, who can know what really happened? Perhaps evolution took place, and perhaps not. It's all "merely a theory" and the creationists do not alter their basic religious beliefs because of an "unproven" theory.

Physical and Spiritual

All these theological problems disappear if one recognizes that the *spiritual* worth of mankind is *unrelated and independent* of his *physical* importance. Indeed, Scripture stresses that spirituality can be found even in the most humble surroundings. The importance of distinguishing between physical and spiritual characteristics is well illustrated by a comparison between chimpanzees and human beings.

In the 1970s, scientific advances yielded a quantitative method for measuring the physical similarities between different species. This new method utilizes molecular biology, and is based on analyzing the thread-like DNA molecules found in every cell of every living creature. The DNA molecule consists of a long chain of elementary units called base-pairs. Studies of the DNA base-pairs have established that the chimpanzee is the primate physically most similar to humans. In fact, the complete array of DNA base-pairs of the chimpanzee is 98.5% identical to that of man, which demonstrates a very close physical similarity between chimpanzees and human beings.[4]

This scientific finding troubles some people, who interpret it as showing that humans are not much different from apes. However, such an interpretation completely misses the point. Anyone can observe that the differences between humans and apes are *enormous*. In the important spiritual realms of creativity, intellect, understanding, and morality, the accomplishments of chimpanzees are totally negligible compared to those of humans. No chimpanzee has ever written a book, painted a picture, proposed a scientific theory, expounded a philosophical thought, or given help to a different species. Indeed, since the physical characteristics of the apes are so very similar to those of human beings, one cannot help wondering why their spiritual/intellectual/creative characteristics are so different. The idea of humans benefiting from divine input suggests itself.

What can one say about man's *physical* capabilities? Humans cannot run like the deer, cannot fly like the bird, cannot swim like the dolphin, cannot climb like the squirrel, cannot chisel like the beaver — the list extends forever. Quite obviously, God did not bestow any special *physical* talents upon mankind. Thus, there is a clear distinction between the spiritual/intellectual/creative and the physical. In the former realms, mankind excels, whereas in the latter realm, we are quite ordinary.

The divine creation described in Genesis maintains this dichotomy between the spiritual and the physical. Therefore, there is nothing degrading about the fact that mankind occupies an ordinary planet that revolves around an ordinary star, and that humans share a common evolutionary history with other animal species. Man's uniqueness, *created in the image of God,* is not expressed in the location of his planet or in the DNA molecules that are found within his cells. Man's uniqueness is to be found in his spiritual/intellectual/creative qualities. It follows, therefore, that there is no need for

---

[4]M. Goodman, 1999, *American Journal of Human Genetics*, vol. 64, pp. 31–39.

the devout person to oppose the heliocentric theory or to denounce evolution.

The scientific discovery of a very ancient universe is also not a problem for the believer. The scientific time scale — billions of years without man — seems so skewed only because of our human perception of time. We are all busy people who would consider it *wasteful* to wait billions of years before getting to the point. But divine considerations are very different. The divine importance of human beings is not to be measured *temporally* by the length of time until our appearance. The crucial point is that *now* we are here, and within an amazingly short time, mankind has assumed the mastery of the physical world, precisely as commanded in Genesis 1:28:

> *"God blessed mankind, and He told them to be fruitful and multiply, and fill the land and conquer it, and rule over the fish of the sea and the birds of the heaven and all the creatures that inhabit the land."*

Chapter 4

Scripture, Science, and Creation III: The Origin of Mankind

Introduction

When did human beings first appear? Scripture tells us (Genesis 1:27) that God created human beings only a few thousand years ago. However, the answer of the scientists is very different.

Scientists assert that hominins ("man-like" species) first appeared about five million years ago. The early hominins then evolved into more advanced hominins, and then into still more advanced hominins, finally culminating in the contemporary human species, Modern Man (*Homo sapiens*). According to scientists, there is no difference between the evolutionary history of man and the evolutionary history of any other species in the animal kingdom.

How is one to resolve this striking contradiction between science and Scripture?

The Difference Between Biology and Scripture

The key to understanding the statement in Scripture that *"God created man in His image"* (Genesis 1:27) is to recognize that Scripture is *not* a biology textbook. Therefore, the word "man" in the biblical creation narrative need not have the same meaning as the word "man" when used by a scientist.

To the biologist, the species Modern Man (*Homo sapiens*), like any other species, is defined by his *physical* characteristics (skull, jaw, teeth, pelvic structure, limbs, etc.) and by his DNA sequences. However, physical features play no role in the biblical understanding of the term "man." Standing two meters tall, walking upright, and possessing a hominin skull are not the relevant criteria in the biblical classification scheme. The "man" in the Genesis creation narrative, who is described as "created by God," is characterized *solely* by his unique creative, intellectual, and spiritual qualities. We shall refer to this creature as "Scripture-man."

A very important feature of "Scripture-man" is his ability to speak. Onkeles, the second-century commentator and translator of the Bible into Aramaic, designates man as "the speaking being." The Jewish commentators Rashi, Sa'adiah Gaon, Sforno, and Nachmanides all emphasize that man's uniqueness lies in his powers of speech and reason.

The origin of human beings is described in Genesis by the words "God created." However, one need not understand these words as implying creation in the physical sense. "Creation" means the formation of something that is *fundamentally new*, either physically (creation *ex nihilo*) or conceptually (the creation of a totally new kind of entity, such as a living creature). Jewish commentators explain that the creation of man "*in the image of God*" refers to the unique intellectual, creative, and spiritual abilities with which man was endowed by his Creator — in other words, Scripture-man.

Scripture-man refers to a creature that displays a very high level of intelligence and creativity, but these features were absent in prehistoric man. It follows that the Scripture account of the creation of man does not refer to prehistoric man.

The preceding discussion suggests that the "creation of man" described in Scripture does not refer to a new species at all, but to sudden and radical changes in human behavior. If these changes in human behavior were so dramatic and revolutionary that they completely altered all aspects of human society, then one can truly say that contemporary mankind was "created" by these changes. This is the meaning of the Genesis words: "God created man."

Have scientists discovered any evidence for sudden, radical changes in human society and in human behavior within the last several thousand years? The remarkable answer to this question is: "Yes."

Archaeological findings show that several thousand years ago, human society suddenly changed so extensively that the scientists speak of a "revolution." Before discussing this revolution, we must first describe the cultural history of prehistoric man.

Prehistoric Man and Modern Man

Neandertal Man

The Neandertals were the prehistoric people who immediately preceded Modern Man. They lived for over 300,000 years throughout Europe, western Asia, and the Middle East. Then, for unknown reasons, the Neandertals suddenly disappear from the fossil record. *Only* Modern Man is found in archaeological sites that date from the last 30,000 years. Names given to peoples who lived during this period, such as Cro-Magnon Man, refer to *cultural* groups, and not to physical types. Cro-Magnon Men were physically as much Modern Men as contemporary Frenchmen or the Chinese.

Scientists agree that Modern Man did *not* evolve from the Neandertals. Ian Tattersall, of the Department of Anthropology of the American Museum of Natural History and a recognized authority on Neandertal Man, writes[1]: "*Homo heidelbergensis quite likely gave rise to the Neandertals, whereas a less specialized population founded the lineage that produced Modern Man.*" Erik Trinkaus, of the University of New Mexico, another authority on the Neandertals, writes[2]: "*Modern humans did not evolve from the Neandertal population.*"

Our principal sources of information regarding the culture of prehistoric man are the tools and other artifacts found in his ancient

---

[1] I. Tattersall, April 1997, *Scientific American*, p. 52.
[2] E. Trinkaus and P. Shipman, 1993, *The Neanderthals* (Jonathan Cape: London), p. 414.

campsites. It might seem natural to ask about the level of sophistication that characterizes prehistoric tools, and to compare them with the tools of Modern Man. However, such a comparison would not be meaningful, because the large well-developed brain of Modern Man gives him an obvious advantage over small-brained earlier hominins. However, there is one exception to the pattern of earlier hominins having smaller, less developed brains. The striking exception is Neandertal Man. According to Trinkaus[3]: *"Neandertals had brains as large and as complex as our own."*

What about the physical capabilities of Neandertal Man? Were these people physically deficient in any way that might have hindered their cultural development? Trinkaus writes[4]:

> *"Neandertals were not less human than Modern Men. They had the same postural abilities, manual dexterity, and range of movement as Modern Men ... a stronger grip than Modern Men, but their control of movement was the same as ours."*

The above discussion shows that it is meaningful to compare the *cultural* achievements of Modern Man with those of Neandertal Man. Cultural differences between these two groups cannot be explained away in terms of physical limitations of Neandertal Man.

Neandertal Culture

What were the tools of Neandertal Man? What were his major artistic achievements? What great cities did he build? What profound writings did he leave for posterity? What important moral teachings did he expound? What marvelous paintings, stirring musical compositions, magnificent sculpture, moving poetry, breathtaking architecture, beautiful gardens, and profound scientific discoveries remain from the Neandertals to mark their 300,000-year-long

---

[3] E. Trinkaus and P. Shipman, 1993, *The Neanderthals* (Jonathan Cape: London), p. 418.

[4] E. Trinkaus and W. W. Howells, December 1979, *Scientific American*, p. 99.

sojourn on our planet? The answer is that their meager cultural legacy embodies *not a single one* of these items!

Scientists have discovered that Neandertal tools were primarily flints with a sharp edge. Their tools look quite similar to the sharp stones that one finds strewn along every beach. In fact, Neandertal tools are so primitive that someone who is not a professional archaeologist would not even recognize them as man-made objects. Tattersall explains[5]:

> "The stoneworking skills of the Neandertals consisted of using a stone core, shaped in a way that a single blow would detach a finished implement. They rarely made tools from other materials. Archaeologists also question the sophistication of their hunting skills. Despite some misleading earlier accounts, no substantial evidence has ever been found for symbolic behavior among the Neandertals or for the production of symbolic objects. Even the Neandertal practice of burying their dead may have been only to discourage hyena incursions, for Neandertal burials lack the 'grave goods' that attest to ritual and belief in an afterlife ... Though successful in the difficult circumstances of the Ice Age, Neandertals lacked the spark of creativity that distinguishes Modern Man."

Regarding artistic accomplishments, it is important to note that the magnificent cave paintings found in southwestern France, Spain, and elsewhere were *all* the work of Modern Man.[6] No cave has ever been discovered that was painted by a Neandertal.

What are the reasons for Neandertal Man's lack of culture? Why was Modern Man able to revolutionize all aspects of his environment, while Neandertal Man hardly left a trace of his existence? In fact, archaeologists must search very hard to find the remnants of Neandertal Man. Recall that the Neandertal brain "does not suggest any differences from Modern Man in intellectual or behavioral capabilities."[7] Scientists have no explanation for the great disparity

---

[5] I. Tattersall, January 2000, *Scientific American*, p. 43.
[6] A. Leroi-Gourhan, June 1982, *Scientific American*, pp. 80–88.
[7] E. Trinkaus and W. W. Howells, December 1979, *Scientific American*, p. 97.

in culture and intellectual achievements between these two hominin species that were so similarly endowed physically.

Modern Man

In discussing the impressive culture that characterizes Modern Man, we need not limit ourselves to the latest technological developments, such as supercomputers and space satellites. As soon as he appeared, Modern Man demonstrated his *enormous* cultural superiority over Neandertal Man. The archaeological data are so striking that every scientific account emphasizes the far-reaching technological advances introduced by Modern Man thousands of years ago. Eric Tattersall writes[8]:

> *"The toolmaking industries of Modern Man are completely different from those of Neandertal Man, reflecting a quantum leap in mental abilities ... Modern Men who followed the Neandertals were their intellectual superiors in every way."*

To summarize, the average brain size and its development are the same for Modern Man and Neandertal Man, and the Neandertals had equal physical capabilities. Therefore, the primitiveness of Neandertal culture is one of the mysteries surrounding these physically quite advanced predecessors of Modern Man.

The Development of Modern Human Culture: The Neolithic Revolution

From his initial appearance, Modern Man gradually developed his technological and artistic skills. Then, about 11,000 years ago, there occurred an explosion of cultural innovations. This was the most comprehensive series of cultural advances that has ever taken place, covering all aspects of human behavior. The cumulative effect of all these changes was to completely revolutionize human society.

---

[8] N. Eldredge and I. Tattersall, 1982, *The Myths of Human Evolution* (Columbia University Press: New York), pp. 154, 159.

Archaeologists refer to the sudden appearance of so many fundamental technological and artistic achievements as the Neolithic
Revolution or the Agricultural Revolution.[9]

The Neolithic Revolution was so all-encompassing that it has
become *the major milestone* in prehistoric chronology. Archaeologists
denote all earlier times as Paleolithic (Old Stone Age), whereas subsequent times are denoted as Neolithic (New Stone Age).

The many fundamental cultural innovations that occurred during or immediately after the Neolithic Revolution include agriculture,
animal husbandry, metalworking, the wheel, first written language,
ceramic pottery, weaving, prepared foods (bread, wine, cheese),
musical instruments, and advanced architecture. This vast prolifera-
tion of cultural advances permitted the formation of complex social
organization that soon gave rise to the first cities and to modern
civilization ("civilization" means "city-making"). The enormous
range of these profound cultural, artistic, and social developments
is emphasized in every archaeological account of this period.
Examples follow:

> *"A crucial event in human history was the beginning of agriculture 11,000
> years ago in the Near East. The accumulation of surplus food supplies
> enabled large settlements to be established, leading to the emergence of
> Western civilization."*[10]

> *"Agriculture and animal husbandry appeared at roughly the same
> time ... Technological progress, the mastery of new materials (such as metals)
> and new energy sources (such as wind and water power) ... The acceleration
> of human history is clearly illustrated by comparing the changes of the past
> 10,000 years with those of the previous four million years."*[11]

> *"One cannot avoid being impressed at how rapidly the transition
> occurred from Paleolithic hunting groups to regionally organized*

---

[9] Strictly speaking, there are differences between the Agricultural Revolution and
the Neolithic Revolution, and these two did not occur at exactly the same time in
each locality. However, such subtleties are of more interest to the professional
archaeologist than to the layman, and we shall here treat these two revolutionary
events as equivalent.

[10] S. Lev-Yadun, A. Gopher, and S. Abbo, June 2000, *Science*, vol. 288, p. 1602.

[11] S. L. Washburn, September 1978, *Scientific American*, p. 154.

communities ... domestication of plants and animals and the establishment of farming communities developed quite rapidly ... Bronze tools were produced. Writing evolved. Market centers became towns ... The urban revolution was underway, the world was radically transformed and the first civilizations were taking shape."[12]

"*The development of plant and animal domestication is referred to as the Neolithic Revolution ... The changes arising from food production so altered human life that all manner of new developments came into being ... village life, population growth, and increasingly complex forms of social organization.*"[13]

"*Major characteristics of agricultural economies began to be evident ... animal domestication and advanced cultivation techniques, such as irrigation, occurred with explosive consequences ... populations increased enormously ... the pace of change was so rapid and their effects were so far-reaching.*"[14]

What were the causes of the Neolithic Revolution? What triggered all these "explosive," "far-reaching," and "revolutionary" changes that so altered human society? The fact is that no one really knows.

"*Why, after hundreds of millennia of subsistence by hunting and gathering, did man recently adopt the alternative strategy of cultivating crops and husbanding animals? The reasons for this dramatic shift remain a topic of debate.*"[15]

The Uniqueness of Mankind

This comprehensive revolution in human society is the true meaning of the Genesis phrase "*God created man.*" The creativity and

---

[12] E. A. Hoebel and T. Weaver, 1979, *Anthropology and the Human Experience* (McGraw-Hill: New York), pp. 183, 195, 201.

[13] G. H. Pelto and P. J. Pelto, 1979, *The Cultural Dimensions of the Human Adventure* (Macmillan: New York), p. 93.

[14] A. Sherratt, ed., 1980, *The Cambridge Encyclopaedia of Archaeology* (Cambridge University Press: Cambridge), p. 407.

[15] L. G. Straus, G. A. Clark, J. Altuna, and J. A. Ortea, June 1980, *Scientific American*, p. 128.

intelligence that suddenly appeared in Modern Man during the Neolithic Revolution remain to this day the distinguishing features of the "man" described in Scripture. The characterization of man as *"created in the image of God"* refers to the unique spiritual, creative and intellectual qualities of contemporary human beings. We shall here discuss three aspects of man's uniqueness.

Language and Communication

The past several thousand years have witnessed enormous progress in all areas of human endeavor. An essential ingredient of this progress is the unique ability of human beings to communicate ideas through speech. This ability enables human beings to benefit from the ideas of others. The distinguished physicist Isaac Newton once remarked: *"If I have seen further than others, it is by standing on the shoulders of giants."*

Human speech should not be confused with the speech of parrots. Unlike parrots, humans have the ability to convey abstract and complicated ideas in many areas.

The importance of the communication of ideas cannot be overestimated. The many technological innovations that have revolutionized human society resulted from the cumulative efforts of many talented people. Because man can communicate ideas, one need not "reinvent the wheel" before making new discoveries. Building upon the work of others has led to the rapid technological progress that is the mark of civilization.

Man's ability to communicate with his fellows is an important aspect of man's having been created *"in the image of God."*

Intellectual Curiosity

Man is the only species that displays intellectual curiosity regarding abstract matters that *do not necessarily enhance his chances for survival.* These include philosophy, art, mathematics, aesthetics, theology, history, science, and psychology. All other species concern themselves only with food, shelter, safety, and mating, for themselves and their family or colony. Human beings express intellectual curiosity and devote much time to the pursuit of knowledge *that has no practical consequences whatsoever*. An excellent illustration of this phenomenon is the book that you are now reading. Reading this book will *not* increase your salary, will *not* put better food on your table, and will *not* improve your physical situation in any way. Nevertheless, in spite of the absence of any practical benefits, you continue to read in order to satisfy your intellectual curiosity.

Man's intellectual curiosity is another aspect of his having been created *"in the image of God."*

Conscience and Morality

The most striking feature of man's uniqueness lies in the realm of conscience and morality. *Only* human beings are capable of making decisions based on the principles of right and wrong. Human beings often sacrifice their personal welfare in the cause of morality. For example, newspaper stories of starving people in Africa generate a worldwide appeal for help. Africans have nothing in common with the average American or European — neither race nor religion nor language nor ideology nor lifestyle. Yet, the sight of starving children touches our hearts, and our conscience demands that we help alleviate the suffering.

Only mankind deals with moral problems. And only mankind possesses the spiritual ability to make moral judgments. This divine privilege and accompanying responsibility are ours alone, because human beings were created *"in the image of God."* As is written in Scripture (Deuteronomy 30:15, 19):

> *"I have set before you this day, life and good, and death and evil ... therefore, choose life."*

Chapter 5

Scripture, Science, and Creation IV: The Wondrous Universe of God

There is a commandment in Scripture (Deuteronomy 6:5) *"to love the Lord, your God, with all your heart and with all your soul and with all your might."* How can one develop love for God, the omnipotent and omniscient Master of the World? The twelfth-century Jewish theologian Moses Maimonides answers this question in his Code of Laws (Laws of the Fundamentals of Belief 2:2):

> *"How can one come to love God? Studying the wonders of nature and God's creatures and seeing in them His infinite wisdom, leads one to love and revere God."*

The idea of studying nature as a means of approaching God was also proposed by Christian theologians, who used the term "natural theology" to denote the study of God through the study of nature. According to the seventeenth-century English theologian Francis Bacon, God wrote two Books: the Book of His words (Scripture) and the Book of His works (Nature).

In this chapter, we shall study the "Book of God's works," meaning, of course, the wondrous universe. A historic account of our growing knowledge of the structure of the universe shows that with each new scientific discovery, an additional layer of wonder is revealed. Indeed, the universe has proved to be far more wondrous than anyone had imagined. These new findings can readily lead one to love and to revere God.

Particles

The universe consists of two components: matter and energy. We shall here discuss matter, whose elementary building blocks are particles. By the eighteenth century, scientists realized that the thousands of different types of materials that are observed are all composed of different combinations of a few basic particles that they called "atoms."

The same atom can appear in very different materials and in very different forms. For example, oxygen atoms are a component of the atmosphere (gas), a component of water (liquid), and also a component of rocks (solid). Although the atmosphere, water, and rocks are very different materials, they all contain the same atoms of oxygen.

Over the years, nearly a hundred different types of atoms were discovered — the atoms of the periodic table. The discovery of these different atoms was an enormous advance in our understanding of nature, reducing the myriads of different materials to only one hundred different atoms.

It had been previously thought that the atom was a basic particle that could not be cut into smaller particles. The very word "atom" indicates this. The first syllable "*a*" means "not," as in "apolitical" or "asexual," and the second syllable "*tom*" means "to cut." Thus, the word "*atom*" means an entity that "*cannot be cut*" into smaller pieces.

By the end of the eighteenth century, it was realized that the tiny atom can be cut into even smaller particles. The picture of the atom that emerged from scientific studies resembles a miniature solar system. The central "sun" of the atom is called the nucleus, around which revolve several "planets." The nucleus of the atom contains two types of particles, protons and neutrons, whereas the role of the "planets" is played by the electrons. Thus, the atom consists of three different types of particles: *protons, neutrons, and electrons.*

Protons have a positive electric charge, electrons have a negative electric charge, and neutrons have no electric charge at all.

Atoms differ from each other according to the number of protons contained in the nucleus. The nucleus of the hydrogen atom contains one proton, the helium nucleus contains two protons, the lithium nucleus contains three protons, and so on, until one reaches the uranium nucleus which contains 92 protons.

The universe of the early twentieth century seemed quite simple, with all matter consisting of only three different kinds of particles: protons, neutrons, and electrons.

The Forces Between Particles

There are forces that act between the particles. These forces are very few. There are two forces that are familiar in our daily lives, which are gravity and the electromagnetic force. It used to be thought that the magnetic force and the electric force were separate forces. However, in 1862, James Clerk Maxwell demonstrated that the electric force and the magnetic force were different aspects of the same force, called the electromagnetic force.

There also exists a third force in nature. The existence of this third force can be demonstrated as follows. The atomic nucleus consists of several protons that are very tightly bound together. Since each proton has positive electric charge, and positive charges repel each other, why doesn't the electromagnetic force split the nucleus apart? What binds the protons together so tightly? The source of the extreme stability of the atomic nucleus must be another force, called the strong nuclear force, that causes protons and neutrons to strongly attract each other. The attractive strong nuclear force between protons is much stronger than the electromagnetic force that repels protons from each other. Therefore, the atomic nucleus is extremely stable.

We have so far discussed three forces: *gravity, the electromagnetic force, and the strong nuclear force.* However, it soon became apparent that this simple picture is not complete. The universe had wondrous surprises in store.

Radioactivity

Early in the twentieth century, the phenomenon of "radioactivity" was discovered. To the surprise of the scientists, it was found that certain atoms are not stable. The nucleus of such atoms spontaneously emits or "radiates" protons and neutrons. An atom having such an unstable nucleus is called "radioactive." Radium and uranium are famous examples of radioactive atoms, whose unstable nucleus spontaneously emits protons and neutrons.

Scientists eventually discovered that the phenomenon of radioactivity is due to an *additional force in nature,* called the *weak nuclear force,* and to an *additional particle of nature,* called the *neutrino.* (The explanation of the connection between radioactivity and the additional particle and the additional force lies far beyond the scope of this book.)

Thus, the universe was more complex, containing *four* different particles (electron, proton, neutron, and neutrino) and *four* different forces (gravity, electromagnetic force, strong nuclear force, and weak nuclear force). However, this was not the end of the story. More surprises awaited.

Quarks

In the middle of the twentieth century, physicists discovered that neither the proton nor the neutron is a fundamental particle. Rather, each is composed of even smaller particles, which are called quarks. Moreover, there are two types of quarks, known as the "up" quark and the "down" quark. Both the proton and the neutron consist of three quarks. The proton consists of two "up" quarks plus one "down" quark, whereas the neutron consists of one "up" quark plus two "down" quarks.

The discovery of quarks did not change the number of basic particles. Previously, one spoke of two particles, proton and neutron, whereas now one still speaks of two particles, "up" quark and "down" quark.

In summary, the universe was now seen to consist of four particles (*"up" quark, "down" quark, electron, and neutrino*) and four forces (*electromagnetic force, gravity, strong nuclear force, and weak nuclear force*). This model of four particles and four forces explained every phenomenon then known to science. Therefore, scientists thought that they had a complete understanding of the universe. However, the universe was soon found to be even more wondrous!

Surprise!

Just when scientists thought they finally understood the universe, another particle was discovered! Upon hearing of this, Nobel laureate Isidor Isaac Rabi asked his famous question: *"Who ordered that?"* In other words: Why does the universe contain a fifth particle? Everything can be understood within the framework of just four particles.

The fifth particle was found to have exactly the same properties as the electron, except that it is about 200 times heavier. It is basically a heavy electron.

More wonders followed. Scientists also discovered *a heavy neutrino, a heavy "up quark, and a heavy "down" quark*, thus yielding *eight* different particles.

By the end of the twentieth century, still more wonders! Scientists also discovered *a still heavier electron, a still heavier neutrino, a still heavier "up" quark, and a still heavier "down" quark*. This yields *twelve* different particles.

To summarize, the universe consists of twelve different types of particles: Three types of electrons, three types of neutrinos, three types of "up" quarks, and three types of "down" quarks. These twelve particles interact with each other by means of four different forces: gravity, electromagnetic force, weak nuclear force, and strong nuclear force. This picture of the universe (twelve particles and four forces) seemed complete.

However, the wonders of the universe were not yet over.

Grand Unification Theory

It has always been part of the scientific enterprise to try to simplify the description of nature by combining the separate forces, showing that they are really different aspects of one single force. Thus, in 1862, James Clerk Maxwell demonstrated that the electric and the magnetic forces are really two aspects of a single "electromagnetic force." He also showed that the phenomenon of light was another aspect of this same force, with light waves being "electromagnetic waves." Thus, electricity, magnetism, and light are not three different phenomena of nature, but only one. This was a great simplification.

In recent years, scientists have continued with this program and have achieved great success. They showed that three of the four forces of nature are different aspects of one single force. The combination of the electric force, the weak nuclear force, and the strong nuclear force is called "Grand Unification Theory." The development of Grand Unification Theory was a crucial advance in our understanding of the universe.

Still More Particles: Higgs Particles and Dark Matter Particles

Among the reasons for its importance, Grand Unification Theory predicts the existence of yet another particle, called the *Higgs particle* (after Scottish physicist Peter Higgs who had proposed it). But there was problem. The Higgs particle had never been detected.

If Higgs particles were discovered, they would serve as a striking confirmation of Grand Unification Theory, the current scientific theory of the forces in the universe. Conversely, if Higgs particles were not discovered, this would call into question the validity of Grand Unification Theory.

In addition to the Higgs particle, scientists discovered, to their great surprise, that the universe must contain still other types of particles. This discovery is connected with what has been called "*dark matter*," where the word "dark" denotes "unknown."

The universe consists of galaxies, which are large clusters of stars. Galaxies rotate because of the gravitational attraction of the stars for each other. The rate of rotation depends on the number of stars in the galaxies. However, a careful study of the rate of rotation showed that the observed galactic rotation is significantly faster than expected.

The rapid rotation of galaxies can only be explained if the galaxy contains much more matter than previously thought. In other words, in addition to its stars, each galaxy must contain additional matter that is invisible to astronomers. Moreover, it has been shown that the particles that comprise this additional matter must be of a type that is *different* from the twelve known particles. This additional matter is called *dark matter*, with the word *dark* denoting both the unknown, *dark* nature of the particles and also that these particles do not *shine* like the particles in stars and hence they are *dark* and invisible to the astronomer.

The most astonishing aspect of dark matter is its sheer quantity. Measurements of the rate of galactic rotation show that dark matter comprises **80% of all the matter in the universe**. Thus, it turns out that for hundreds of years, scientists have been studying only a minor component of the mass of the universe. They were totally unaware of the existence of the major component of the mass of the universe — dark matter. The wonders of the universe never cease!

Thus, we see that there are two types of particles that scientists must detect: Higgs particles and the particles that comprise dark matter.

Producing Particles

Scientists began to seek these particles by means of an instrument called an accelerator, which is capable of producing particles. In order to produce Higgs particles and dark matter particles, in 2012, scientists constructed a giant accelerator in Geneva, which is known technically as the "large hadron collider." It was hoped that the giant Geneva accelerator would produce both Higgs particles and the particles of dark matter.

An accelerator produces particles in the following way. When an electrically charged particle, such as a proton, is placed within an electric field, the proton is accelerated. If the proton is constrained to move in a circular orbit (which can be arranged by means of magnets), then every time its circular orbit brings the proton back into the electric field, the proton will receive another "push" and thus will go even faster.

The accelerator works like a person pushing a swing. The swing is pushed forward each time it returns to the "pusher." But unlike a proton in an accelerator, the swing does not go faster with each push because friction slows it down. In an accelerator, the protons move in a vacuum where there is no friction. Therefore, with each "push" by the electric field, the protons move still faster. Many protons that are accelerated together form a "beam" of speeding protons.

A moving particle has kinetic energy, the energy of motion. Therefore, the beam of rapidly speeding protons has an enormous amount of kinetic energy. In the accelerator, two such beams of protons are formed, with the beams aligned to move in exactly opposite directions, aimed to collide head-on. For this reason, scientists often refer to an accelerator as a "collider."

When two speeding cars collide head-on, they come to a stop, and thus lose all their kinetic energy. The supposedly "lost" kinetic energy goes into smashing up the cars. Also in an accelerator, when the two proton beams collide, the protons stop. But unlike the colliding cars, the supposedly "lost" kinetic energy from the proton beams produces new particles.

The production of particles from kinetic energy is described by Einstein's famous equation: $E = Mc^2$, where the letter E denotes energy, M denotes matter (in the form of particles), and c denotes the speed of fight. Einstein's equation states that it is possible to convert matter (M) into energy (E). This is the basis for nuclear energy, for nuclear bombs, and for nuclear power stations.

Einstein's formula works in both directions. Not only is it possible to convert matter into energy, it is also possible to convert energy into matter. This is what the accelerator does. It converts the enormous "lost" kinetic energy of the speeding proton beams into matter in the form of particles.

The more kinetic energy that is available, the larger will be the mass of the newly formed particle. In terms of Einstein's equation, more E will yield more M. Therefore, to produce a very heavy particle (say, more than 100 times the mass of a proton), one must have a very large E. That is, one must accelerate the protons to an extremely high kinetic energy. In order to produce extremely fast, high-energy protons, the accelerator must generate huge electric fields which must be perfectly aligned. This is one of the reasons that the Geneva accelerator is so complicated.

The giant Geneva accelerator consists of an enormous underground tunnel that is about 20 meters in diameter and 27 kilometers in length. This tunnel is crammed with complicated scientific instruments. The giant Geneva accelerator is the largest and most expensive scientific facility ever built.

According to Grand Unification Theory, the Higgs particle is too heavy to have been produced by existing accelerators. Therefore, it is understood why this particle had not been detected previously. However, the Geneva accelerator is powerful enough to produce Higgs particles. This was one of the primary tasks of the Geneva accelerator. Accordingly, in 2012, it was announced in Geneva that scientists had finally obtained definitive evidence for the existence of Higgs particles. This discovery, which served to confirm Grand Unification Theory, was considered so important that it was reported internationally and resulted in the award of the Nobel Prize.

Dark matter particles must be very heavy or they would have been produced by existing accelerators. It is hoped that the powerful Geneva accelerator will also produce the mysterious particles of dark matter, and thus shed light on their nature.

Theory of Quantum Gravity

The modern theory for understanding the universe is called quantum theory. Therefore, every theory of physics must be compatible with quantum theory. Grand Unification Theory is compatible with quantum theory, but the theory of gravity *is not*. Therefore, the current theory of gravity cannot be correct. Scientists struggled for many years, but without success, to formulate a new theory of gravity that would be compatible with quantum theory — a theory called *quantum gravity*.

In the 1990s, there was an important breakthrough. Scientists finally succeeded in formulating a theory of quantum gravity, by means of a new concept of the universe known as "*string theory*."

String Theory

The central idea of string theory is quite wondrous. According to string theory, the basic entities of the universe are not tiny particles (such as electrons or quarks), as previously thought. Rather, the basic entities of the universe are tiny strings of infinitesimal length. These tiny strings vibrate and the vibration of the strings expresses itself as energy. According to Einstein's equation, $E = Mc^2$, energy is equivalent to matter. Therefore, the concentrated energy of the tiny vibrating strings appears to us as tiny particles of matter, in spite of the fact that they are not, in fact, particles.

Note what has happened. Our entire concept of matter has been radically altered! According to string theory, *tiny vibrating strings* are the fundamental entities of matter.

String theory makes even more wondrous predictions. One such prediction is that the universe does not consist of only the three familiar dimensions (up-down, right-left, and forward-backward). Rather, **the universe consists of ten dimensions**, the familiar three dimensions mentioned above plus an additional seven dimensions that are too tiny to be observed. These extra dimensions are so small that it is impossible to detect them. If so, why do scientists claim that these extra dimensions exist? The reason is that the existence of a quantum theory of gravity *requires the existence* of a multi-dimensional universe. Although the dimensions of the universe are a very interesting and very important topic, the multi-dimensional character of the universe will not be discussed further.

It must be admitted that it is still not completely clear that string theory is the correct description of nature. However, the evidence is piling up that string theory in its modern form (known technically as *M*-theory) is indeed the correct theory of nature.

The feature of string theory that is relevant to our discussion is that string theory predicts the existence of additional string particles, which are of a totally different nature from any other known particle. However, the string particles predicted by string theory have never been observed. The production of the predicted string particles by the Geneva accelerator would constitute overwhelming evidence for the correctness of string theory.

Scripture and the Wondrous World of God

In recent years, leading scientists have emphasized that the universe has proved to be much more wondrous than expected. For example, physicist Brian Greene, of Columbia University, has written a book on this subject, entitled *The Elegant Universe*. In the course of 450 pages, Greene exposes the reader to the many wondrous aspects of the universe.

The world is so wondrous! It extends from the vastness of outer space with its countless galaxies to the microscopic realm of tiny string particles. The glory of God is seen everywhere in the universe! According to Moses Maimonides, quoted earlier, "*attaining insight into God's wondrous universe provides a glimpse into His infinite wisdom and enables one to develop a feeling of awe and love for the Master of the Universe.*" As we continue to learn more about the wonders of the universe, our awe of God can only increase.

A description of the wonders of nature and their relationship to God is found throughout Scripture. Examples follow:

- In the Book of Nehemiah, it is written (9:6): "*You are the Lord Who has made the heaven with all its host, the earth and all the things that are upon it, the seas and all that is in them.*"
- In the Ten Commandments, the following reason is given for observing the Sabbath (Exodus 20:11): "*Because ... the Lord made*

the heaven, the earth, the seas, and all that they contain, and He rested on the seventh day."

- In Psalms, it is written: (19:2–5): "*The heaven proclaims the glory of the Lord, the sky declares His handiwork. Day after day pours forth speech, and night after night reveals knowledge. There are no words. Unheard is their voice. Yet their message extends throughout the earth and their words reach the end of the world.*" This passage states that there is no need for words to explain the greatness of God. The wonders of the universe speak for themselves.

- In Psalms, it is written: (33:6–8): "*By the word of God, the heaven was made, and all its host by the breath of His mouth. He gathers the water of the sea in a heap. He places the deep in storehouses. Let all the earth revere the Lord. Let all the inhabitants of the world stand in awe of Him.*"

- In Psalms, it is written: (95:1–5): "*Let us sing to the Lord ... Let us acclaim Him with songs of praise ... In His hand are the depths of the earth. The mountain tops are His. His is the sea, for He has made it. His hands formed the dry land.*"

- In Psalms, it is written: (Chapter 136): "*He performs great wonders ... He made the heaven with wisdom ... He spread the earth over the water ... He made the great luminaries ... the sun to rule by day, the moon and the stars to rule by night.*"

- In the Book of Isaiah, it is written (40:21–22, 26): "*Do you not know? ...Have you not understood the foundations of the earth? ... He (God) stretches out the heaven as a curtain and spreads it out as a tent ... Lift up your eyes and you will see Who has created all these ... He brings out the Host by the greatness of his power.*"

"How wondrous are Your deeds, O Lord,
You have made everything with wisdom."
(Psalms 92:6)

Chapter 6

The Second Creation Narrative in Scripture

Scripture begins by describing how God created the world. The importance of the creation narrative has been emphasized by the Jewish commentator Moses Nachmanides (thirteenth century), who writes:

> *"The belief that God created the world is the essence of our faith. One who does not believe this is a heretic."*

The creation narrative in the first chapter of Genesis begins with the creation of the world (*"In the beginning, God created the heaven and the earth"*) and ends with the creation of man (*"God created man in His image"*). Surprisingly, however, this creation narrative is immediately followed by a second creation narrative in the second chapter of Genesis.

The Second-Chapter Creation Narrative

The second-chapter creation narrative is given below (Genesis 2: 4–7). For the Hebrew word י-ה-ו-ה (Tetragrammaton), we shall use the term "Hashem" ("The Name"), as is commonly done in Jewish writings. For the Hebrew word אלהים (Elohim), which denotes power and strength, we shall use the word "God."

Verse 2:4: *"These are the generations of heaven and earth when they were created, on the day that Hashem God made earth and heaven."*
Verse 2:5: *"No tree was yet on the earth and no vegetation had yet grown because Hashem God had not brought rain, and there was no man to till the soil."*
Verse 2:6: *"A mist ascended from the earth and watered the surface of the ground."*
Verse 2:7: *"Hashem God formed man from the dust of the earth and He breathed into his nostrils the soul of life, and man became a living being."*

Various approaches have been suggested to explain the two creation narratives in Genesis. We will here present an approach that answers the following questions:

• What does the second-chapter creation narrative add to what is already written in the first-chapter creation narrative?
• Why are different expressions used to designate God in the two narratives? In the first chapter, God is called "Elohim" (*"Elohim created ..."*), whereas in the second chapter, God is called "Hashem Elohim" (*"Hashem Elohim formed ..."*).

The second-chapter creation narrative consists of only four verses (2:4–7). And these verses contain no mention of light or animals or oceans or dry land or heavenly bodies.

Verse 2:4 is introductory. Verse 2:5 discusses the *absence* of vegetation, which is explained by the fact that God had not yet provided the necessary rain and man had not yet appeared to till the soil. The next two verses (2:6–7) discuss the appearance of the "missing" items. God provides the necessary water (2:6) and man appears to till the soil (2:7). Therefore, vegetation can now sprout. This is the entire description of the creation. Is there nothing else important to mention in the second creation narrative?

Verse 2:5 raises another question. There was no need for man to *"till the soil"* to produce the dense jungles of South America and Africa or the vast forests of Europe. Therefore, it is clear that the vegetation being discussed refers to agricultural produce — grain, fruits, and vegetables — which does require man to *"till the soil."*

The Partnership Between Man and God

The thesis proposed here is that the second-chapter "creation narrative" is not a creation narrative at all. Rather, Verses 2:5–7 illustrate an important principle of Scripture, namely, *every creative activity of mankind is based on a partnership between man and God.* This partnership is demonstrated in the second chapter by agricultural produce, which requires *both* an act of God (to supply the rain) *and* an act of man (to till the soil). Verses 2:5–7 thus *complement* the first chapter of Genesis, with both chapters *together* forming a *single creation narrative.*

Consider the following examples of the partnership between man and God.

Man	Developed by Man	Supplied by God
Farmer	grain, fruits, vegetables	water, soil
Potter	pottery	clay
Smith	metallic implements	metallic ores
Jeweler	jewelry	gold, silver, precious stones
Artist	paintings	pigments
Physician	medicines	chemicals

The Talmudic Sage Rabbi Akiva illustrated the idea of the partnership between man and God by means of a loaf of bread. Rabbi Akiva pointed out that the loaf of bread produced by the baker (man's contribution) is even more impressive than the grains of wheat (God's contribution) from which the bread was baked. Thus, even an item as simple as a loaf of bread requires a partnership between man and God, with man's contribution being quite important.

The Names of God

In the first chapter of Genesis, the term designating God is Elohim (in Hebrew: אלהים), whereas in the second chapter, Hashem Elohim (in Hebrew: אלהים י-ה-ו-ה) designates God. Why is Hashem ("The Name") added to denote God in the second chapter?

Different names for God appear in the first two chapters of Genesis because of the different contents of these two chapters.

The term Elohim denotes power. The term Elohim is also used in Scripture to refer to the god (power) worshipped by other nations. For example, in the Ten Commandments, God states: "*You shall have no other god (Elohim) before Me*" (Exodus 20:3).

The first chapter emphasizes the *power* of God, describing how He created the entire world. Therefore, the appropriate term for God is Elohim. However, the subject of the second chapter is very different. This chapter deals with the *partnership between man and his Creator*. Hence, in the second chapter, the *personal* name of God, namely, Hashem, has been added and God is referred to as Hashem Elohim.

The Appearance of Mankind

The first chapter of Genesis describes the appearance of man. However, the appearance of man is again described in the second chapter of Genesis. The two descriptions are:

Verse 1:27: "*God **created** man in His image, in the image of God He created him. Male and female He created them.*"
Verse 2:7: "*Hashem God **formed** man from the dust of the earth, and He breathed into his nostrils the soul of life, and man became a living creature.*"

These two verses raise the following questions:

- Why are two very different verbs used to describe the act of God that led to man? "*created*" (in Hebrew: *bara*) in the first chapter but "*formed*" (in Hebrew: *yatzar*) in the second chapter.
- Why does the second-chapter narrative state the material from which man was formed ("*the dust of the earth*")? What can be learned from this information?

The first part of Verse 2:7 describes the *physical* nature of man ("*the dust of the earth*"), whereas the second part of the verse describes the *spiritual/creative/intellectual* nature of man ("*the soul of life*").

Verse 2:7 thus describes the **dual nature of mankind** — *both* physical (*dust*) *and* spiritual/creative/intellectual (*soul*). The combination of physical *and* spiritual constitutes the essence of mankind and sets him apart from all other creatures.

The unique feature of man lies in his spirituality. Therefore, Scripture describes man in Verse 1:27 as being *"created in the image of God."* Accordingly, all traditional Jewish commentators describe man in terms of his creative and intellectual capabilities.

- Onkeles: *"Man is a speaking creature."*
- Rashi: *"His body is from the lower world, but his soul is from the upper world ... knowledge and the power of speech were given to him."*
- Ramban: *"He has knowledge and intelligence ... his soul enables him to speak."*
- Sforno: *"Man was unable to speak until he was created in His image."*

In view of the above, one can understand why two different verbs were used to describe the appearance of man. Verse (1:27) describes the unique *spiritual/creative/intellectual* powers of man. Since these aspects of mankind are fundamentally new to the animal kingdom, the appropriate verb is *"created."* By contrast, the first part of Verse 2:7 describes the *physical* aspects of man ("the dust of the earth"), which mankind shares with all other creatures. Therefore, the verb *"created"* would be inappropriate and the verb *"formed"* is used.

The Uniqueness of Mankind

Are there any signs that human beings possess unique spiritual qualities? In fact, nothing is more obvious than the *uniqueness* of the enormous intellectual and creative abilities of human beings. This can be illustrated in a very interesting way by the work of primatologist Sue Savage-Rumbaugh. After considerable effort, she succeeded in teaching a bonobo chimpanzee (the species most similar to humans) named Kanzi to recognize as many words as are learned, *completely effortlessly*, by a human child of two and a half years.

This "intellectual achievement" on the part of the chimpanzee serves to emphasize the vast chasm that separates the mental capabilities of man from those of *every other species.*

The prestigious journal *Scientific American* devoted an entire issue (September 2018) to describing the unique features of human beings. On the cover of that issue are emblazoned the seminal words: "HUMANS: Why we're unlike any other species on the planet."

There are those who vigorously deny the uniqueness of human beings, insisting that humans are just another one of the two million species of animals thus far identified. This point of view admits that we are different from other species, but every species possesses some special properties that set it aside as a separate species. The claim is that it is only human pride that makes us think that we are unique creatures who were "*created in the image of God.*"

Our lack of uniqueness is the thesis of Jared Diamond's book, "*The Third Chimpanzee.*" The title refers to human beings, whereas the other two species are the common chimpanzee and the bonobo chimpanzee. Diamond asserts that human beings have no special talents that are not shared, to some extent, by many other animals. He claims that even our ability to think does not fundamentally distinguish us from the other species of chimpanzees.

In fact, the uniqueness of human beings is blatantly obvious and can easily be demonstrated. Consider the following facts about the University of California at Los Angeles (UCLA), Professor Diamond's home university. A Google search reveals that UCLA has a student enrollment of 40,000 and their libraries contain over eight million volumes. There is something very important about these data. *Not a single one of the 40,000 UCLA students is a chimpanzee!* And there is something else equally important. *Not a single one of the eight million books in the UCLA libraries was written by a chimpanzee!*

The above facts are especially surprising because of the close *physical* similarity between the two species. Diamond points out that 98.4% of the genetic material of humans is also found in chimpanzees. Since genes determine the physical properties of an animal, this close genetic similarity shows that *physically,* we are very similar to the chimpanzee. This fact immediately raises the following important question. If we are so very similar to the chimpanzee *physically,* why are we so different *intellectually and culturally?* One possible answer is that we have been blessed with divine input.

The *spiritual/creative/intellectual* powers of contemporary human beings are also vastly superior to those of prehistoric men. No prehistoric man ever established a university, ever wrote a book, or ever displayed any other feature of civilization. Prehistoric men first appeared five million years ago. That is plenty of time to have accomplished something. One might attribute the lack of cultural accomplishments of the chimpanzee to his smaller brain size, but this argument does not apply to the prehistoric man known as Neandertal Man, *whose brain size was equal to that of Modern Man.*

The same situation also applies to Modern Man (*Homo sapiens*), our own species. Modern Man first appeared about 200,000 years ago. However, the earliest signs of civilization occurred only a few thousand years ago. What took so long? In fact, the appearance of civilization was so sudden and so dramatic that the archaeologists speak of a revolution in human behavior — the Neolithic Revolution — whose causes remain a mystery to this day.

The archeological data strongly suggest that human beings are fundamentally different from all other creatures. One of the most interesting aspects of the many unique features of human behavior is that they manifested themselves quite *suddenly* only a few thousand years ago. Archaeological sites testify to the recent appearance of civilization.

The sudden recent appearance of civilization is incompatible with the scenario of Diamond, but it is in harmony with the words of Scripture (Genesis 1:28): *"God blessed mankind, commanding him to be fruitful and multiply, to fill the land and to subdue it."*

Chapter 7

Scripture and the Heavenly Bodies I: The Stars

The Question

Scripture relates (Genesis 1:14–18) that God caused the heavenly bodies to appear in the sky for the benefit of mankind. These include the sun, moon, and stars. The purpose of the sun is clear. Indeed, without the warmth and energy of the sun, life on earth would be impossible. But what benefit does mankind derive from the moon and the stars? They seem to be mere attractive ornaments that decorate the night sky. In the next two chapters, we shall show how important to mankind are the stars and the moon. In this chapter, we deal with the stars, and in the following chapter, we shall deal with the moon.

People have always enjoyed looking at the stars. They are a delight to the eyes as they twinkle in the sky on a clear night. With the naked eye, one can see only a limited number of stars, perhaps a few thousand. However, it has now recognized that the actual number of stars is not limited to a few thousand. With the aid of powerful telescopes and other modern instruments, astronomers realize that there are untold billions of stars in the sky. Stars cluster into galaxies — our galaxy is called the "Milky Way" — each of which contains *many billions of stars*. And there are *billions of galaxies in the universe.*

In past years, many people thought that the distant stars have some beneficial effect on mankind. However, this idea is now recognized to be astrology and superstition, long abandoned by intelligent people. Therefore, since the many billions of stars seem to serve no purpose, what is meant by the statement in Scripture that God caused the stars to appear in the sky for the benefit of mankind?

The Distant Stars and Life on Earth

As unexpected as this may seem, recent scientific discoveries demonstrate that without the distant stars, life on earth would have been impossible.

The bodies of all living creatures are composed of various chemical elements, including carbon, oxygen, nitrogen, and smaller but vital amounts of other elements. What is the origin of these chemical elements?

According to the firmly established Big Bang theory of cosmology, the universe began with the sudden appearance of an enormous ball of light. Scientists refer to this as the primeval light-ball, but it also has the popular name of the big bang, and hence the name of the theory. Most of the energy of this light-ball was transformed into matter according to Einstein's famous formula, $E = Mc^2$, where the letter E denotes energy (light is a form of energy) and the letter M denotes matter. Not only can matter be transformed into energy, which is the source of nuclear energy, but energy can also be transformed into matter. However, under the conditions prevalent at the time of the big bang, the only matter that was formed in the universe consisted of the simplest elements hydrogen and helium. There was neither carbon nor oxygen nor nitrogen, nor any of the other elements that are vital to life. Where did these more complex elements come from?

This was a puzzle for a long time, until resolved by Fred Hoyle, of the University of Cambridge. Hoyle demonstrated that the stars were the source of all chemical elements with the exception of hydrogen and helium. Inside a star, the temperature is so high and the pressure is so great that the simple elements hydrogen and helium are fused together to form the more complex chemical elements that are vital for life.[1]

However, it does not seem beneficial to mankind that the raw materials of our bodies are formed in the interior of stars. One cannot live inside a star. However, a remarkable event sometimes happens. Occasionally, an entire star explodes, an event called a supernova explosion, which causes the chemical elements from inside the exploding star to be ejected into space. There, the elements then gather to form planets, rocks, air, water, and the tissues of our bodies.

This is explained in *The Cambridge Encyclopedia of Astronomy*[2]:

> *"It used to be thought that the universe has always had the composition that we observe today ... It is not widely appreciated that all the atoms on Earth (except hydrogen and helium) were created inside a generation of stars that evolved before the birth of our Sun and planets. All these elements were made in stellar furnaces before our Sun and planets were formed. The elucidation of the process by which elements were formed in the cosmic environment will stand in history as one of the greatest advances in the twentieth century."*

The article concludes with a sentence that is rarely found in a scientific encyclopedia:

> *"In truth, we are the children of the universe."*

Just as a child cannot come into existence without parents, in the same way, human beings could not have come into existence without the entire universe with its billions of stars. The idea that the entire universe was necessary for our existence is exactly what is written in Scripture.

---

[1] M. Zeilik, April 1978, *Scientific American*, p. 110.
[2] S. Mitton, 1977, *The Cambridge Encyclopedia of Astronomy* (Jonathan Cape: London), pp. 121, 123, 125.

These words — *"we are the children of the universe"* — do not appear in any theological treatise. But they do appear in *The Cambridge Encyclopedia of Astronomy*. Science and Scripture have indeed merged.

The Vast Distances Between Us and the Stars

There is another important connection between life on earth and the stars. This relates to the vast distances that separate us from the stars. Even the nearest stars are more than four light years distant. (A light year is the distance that light travels in an entire year, over 15 trillion kilometers.) And this is just the distance to the nearest stars, with most stars being vastly more distant from us. Why is this crucial, for life on earth?

Supernova explosions emit not only the chemical elements that are essential for life but they also emit "cosmic radiation," which is deadly. We are saved from cosmic radiation *only* because the stars are so distant from us. Over the enormous distances that cosmic radiation travels before reaching the earth, the intensity of the deadly radiation becomes so reduced that it is no longer harmful. Freeman Dyson, of the Institute for Advanced Study in Princeton, explains[3]:

> *"The vastness of interstellar space dilutes cosmic rays enough to save us from being fried, or at least sterilized, by them. If sheer distance had not effectively isolated the quiet regions of the universe from the noisy regions, no type of biological system would have been possible."*

Professor Dyson also writes[4]:

> *"As we look out into the universe and identify the many accidents of physics and astronomy that have worked together for our benefit, it almost seems as if the universe must, in some sense, have known that we were coming."*

---

[3]F. Dyson, September 1971, *Scientific American*, p. 57.
[4]F. Dyson, September 1971, *Scientific American*, p. 59.

One notes the harmony between these words of a world famous secular scientist and the words of Scripture with which this chapter began.

Summary

One sees that not only were the distant stars necessary for life to appear on earth but the distant placement of the stars was also crucial for life to exist. The various details of star formation merge to permit us to live and flourish on our home planet, exactly as stated in Scripture.

Chapter 8

Scripture and the Heavenly Bodies II: The Moon

Scripture tells us that God placed the heavenly bodies in the sky for the benefit of mankind. In the last chapter, we dealt with the importance of the stars. In this chapter, we shall discuss the importance to mankind of the moon.

Questions

According to Scripture (Genesis 1:14): "*the heavenly bodies [the sun and the moon] ... determine the seasons, the days, and the years.*" It is clear that the sun performs these functions, but why is the moon mentioned? It seems that the moon has no effect on the length of the day or the year or the seasons, which are determined solely by the sun. Therefore, how can one understand the statement in Scripture regarding the moon?

The Influence of the Moon

Modern science has made remarkable discoveries regarding the influence of the moon on the earth. In the following pages, we shall describe in detail the reasons for this influence. It is useful to first summarize the importance of the moon.

The moon has played a crucial role in producing and maintaining the present mild climate on the earth. Indeed, the pleasant weather that we enjoy today is directly related to the influence of the moon on the earth. The present-day weather is completely different from the extremely harsh conditions that prevailed in the past. The mild climate of the past few thousand years has permitted the flourishing of human society.

Major technological advances have transformed mankind from a modest inhabitant of the earth to a triumphant conqueror of outer space — all within the space of a few thousand years. This unprecedented blossoming of human activity, leading to accelerated cultural development, has been made possible by the recent dramatic improvement in the weather worldwide.

We now present the scientific evidence that explains the influence of the moon on the earth.

The Day, the Year, and the Seasons

Few statements seem more firmly established than the assertion that the day has 24 hours, that the year has about 365 days, that a compass needle always points "North," and that in the Northern Hemisphere, the summer begins in June and the winter in December. But it was not always so.

Geological evidence has established that in the Devonian Period (450 million years ago), the day had only 21 hours and the year had more than 400 days.[1] Moreover, studies of magnetic sediments on the ocean floor have shown that a million years ago, compass needles pointed South.[2] Finally, the Northern Hemisphere summer once started in December and the winter began in June.[3]

---

[1] J. Andouze *et al.* (eds.), 1985, *The Cambridge Encyclopedia of Astronomy* (Cambridge University Press: Cambridge), p. 54.

[2] D. G. Smith (ed.), 1981, *The Cambridge Encyclopedia of Earth Science* (Cambridge University Press: Cambridge), p. 120.

[3] J. G. Lookwood, 1985, *World Climate Systems* (Edward Arnold: London), p. 110.

If the solar system consisted of only the sun and the earth, then neither the length of the day nor the number of days in a year would ever change, and June would always herald the onset of summer in the Northern Hemisphere. In fact, however, the earth is also subject to the gravitational attraction of the moon. (The gravitational attraction of the other planets is slight.) As a result, the motion of the earth is not limited to its primary motions of rotating about its axis and revolving around the sun. The gravitational attraction of the moon causes the earth to undergo important secondary motions. It is precisely these complex secondary motions that are responsible for the changes in the day, the year, and the seasons.

The gradual motion of the earth that is caused by the moon may seem to the layman to be a mere esoteric detail, no doubt interesting to the professional astronomer, but of no great practical significance. We shall show how incorrect such an assessment is. In fact, it is not an exaggeration to say that the very fabric of human existence has been altered by the seemingly minor changes that the moon causes on the motion of the earth.

The Moon and the Ice Ages

The earth is a warm, hospitable planet, whose mild climate is conducive to man's physical well-being and cultural development. But it was not always so. Only 18,000 years ago, our planet was caught in the grip of a savage ice age. Fully one-third of the land area lay buried beneath massive glaciers whose thickness was measured in kilometers.[4] Glaciers extended so far south that the ground beneath the present-day cities of New York, Paris, London, and Berlin was covered by thick sheets of ice.

About 10,000 years ago, the ice age began to subside. The glaciers retreated, global temperatures began to rise, and the ice sheets melted. In the course of this warming, a milder climate gradually spread over most of the globe. Meteorological evidence shows that

---

[4] C. Covey, February 1984, *Scientific American*, pp. 42–50.

as the land was freed from the ice, profound worldwide changes occurred in the weather.[5]

> *"A renewed rise in temperature set in at 10,000 BP (BP = years before present) which led to the sustained warm climate of postglacial times ... during the period 9000-8000 BP, temperatures continued to rise ... the colder seasons of the year became milder ... summers gradually became warmer."*

The fierce ice age was over, and the world entered the warm interglacial period that continues to the present day.

The changes in society that occurred after this vicious ice age ended were so striking that scientists have given the current interglacial period a special name — the Holocene Epoch. The emphasis of this name is *not* on the marked change in climate. Rather, the name indicates the extensive evidence for man's accelerated cultural development. *The Cambridge Encyclopedia of Archaeology* explains[6]:

> *"The latest warm interval, which began about 10,000 years ago and is still continuing, is conventionally referred to as the Holocene ... The distinction, emphasizing the uniqueness of the last 10,000 years, is more important in terms of cultural development than of climatic patterns."*

The Milankovitch Theory of Ice Ages

The glaciation of 18,000 years ago and the subsequent melting of the glaciers starting about 10,000 years ago marked the most recent of the ten ice ages that have occurred during the past million years.[7]

> *"A growing body of evidence strongly supports the idea that ice ages are caused by small changes in the tilt of the earth's axis and in the geometry of the earth's orbit around the sun."*

---

[5] H. H. Lamb, 1977, *Climate: Present, Past, and Future* (Methuen: London), pp. 371–372.

[6] A. Sherrant (ed.), 1980, *The Cambridge Encyclopedia of Archaeology* (Cambridge University Press: Cambridge), p. 52.

[7] C. Covey, February 1984, *Scientific American*, p. 42.

This conclusion is based on the following evidence[8]:

"Studies of sedimentary deposits in the oceans have shown that astronomical variations played a key role in causing these recent dramatic changes in climate."

The Serbian astronomer Milutin Milankovitch was the first to propose that the small changes in the motion of the earth are the cause of the recurrent cycle of an ice age followed by a warm interglacial period. The Milankovitch theory relates the world's climate to three parameters that characterize the earth's orbit: (1) the degree to which the elliptical orbit departs from a perfect circle, called the "eccentricity," (2) the angle between the earth's axis of rotation and the direction perpendicular to the plane of its orbit, called the "axial-obliquity," and (3) the direction in which the axis points when the earth is closest to the sun, called the "season of perihelion." In 1976, decisive evidence in favor of the Milankovitch theory was established.[9]

We close the discussion by emphasizing that each of the three orbit parameters changes as a result of the small, but important gravitational pull of the moon on the earth.

The Uniqueness of Our Moon

It is particularly interesting to note that our moon is unique among the many moons that surround the planets in our solar system. The uniqueness lies in the fact that our moon is *huge*. The intention here is not huge in absolute size, because the large moons of Jupiter and Saturn are larger than our moon. The intention is that *the moon is huge relative to the size of its mother planet*. The diameter of the largest moon of Jupiter (Ganymede) and the largest moon of Saturn (Titan) is only 4% of the diameter of their mother planet. Therefore,

---

[8] J. K. Beatty *et al.*, 1981, *New Solar System* (Cambridge University Press: Cambridge), p. 69.

[9] J. G. Lookwood, 1985, *World Climate Systems* (Edward Arnold: London), p. 110; C. Covey, February 1984, *Scientific American*, p. 47.

these moons are far too small to affect the orbit of Jupiter or of Saturn. By contrast, the diameter of our moon is 27% of the diameter of the earth, large enough to critically change the orbit of the earth.

The reason for the huge size of our moon is directly related to the manner in which the moon was formed. Unlike all other moons in the solar system, our moon was formed through a collision between a Mars-sized astronomical object and the earth. The resulting violent collision ejected into space vast quantities of material from the earth, and this material eventually coalesced to form our moon.[10]

Summary

The moon has played a central role in fixing the orbital parameters of the earth. Moreover, according to the Milankovitch theory, the gravitational attraction of the moon on the earth is also responsible for the present-day mild climate which has been so conducive to man's unprecedented cultural, intellectual, and technological development. Therefore, the verse in Scripture that lists the moon among the astronomical bodies that determine *"the seasons, days, and years"* is consistent with current scientific knowledge.

---

[10] R. Canup and E. Asphaug, 2001, *Nature*, vol. 412, pp. 708–712.

Chapter 9

Scripture and the Heavenly Bodies III: The Sun

The Genesis creation narrative in Scripture presents questions that have been the subject of many articles and books. However, there is one question that has received relatively little attention and this is the question that we will address here.

The Source of the Food That We Eat and the Oxygen That We Breathe

All the food that we eat that keeps us alive comes from plants. Even when we eat meat, the animals that we eat obtained their food from the plants they ate or from other animals that in turn got their food from plants. Without plants, all living animals would soon die.

Plants make their food by a process called photosynthesis. In this process, plants absorb water from the ground and carbon dioxide from the air and combine these two molecules to form oxygen and energy-rich organic compounds — sugar, proteins, fats — that constitute our food. The oxygen thus produced escapes into the atmosphere, which we then breathe. However, this conversion process requires chemical energy. What is the source of this chemical energy?

Green plants have molecules in their leaves called chlorophyll. These complex molecules are able to absorb sunlight and to transform the energy of sunlight into chemical energy. This chemical energy, produced by sunshine acting on chlorophyll, enables plants to produce the food that we eat and the oxygen that we breathe. Since food and oxygen are both necessary for life, if there were no sunshine, there could be no life.

Question

What does Scripture say about plants and the sun? The Genesis creation narrative states that the plants appeared on the Third Day of Creation (*"God said, 'Let the earth produce vegetation, seed-bearing plants and trees'... and so it was."*). It also states that the sun appeared on the Fourth Day of Creation (*"God made two great luminaries, the greater one (sun) to rule by day ... God set them in the heaven to shed light upon the earth."*).

Scripture thus states that the plants appeared *before* the sun. But this is not possible because plants cannot exist without sunlight. Therefore, how can one understand these verses of Scripture? The answer is based on an important principle relating to the words of Scripture, which we shall now discuss.

An Important Principle

The development of the universe into its present form was an extremely complex process, involving many intermediate stages that occurred over long periods of time. Books on cosmology, astronomy, geology, and biology describe these intermediate stages in great detail. However, Scripture is not a science book. Its purpose is to emphasize to mankind how wondrous is the universe that *we now observe*. The intermediate stages that were necessary to produce this wondrous universe of today are no longer visible and have only recently been discovered by scientists. However, Scripture is intended for everyone, including both the people who lived in the ancient world, as well as the professional scientists of today. Therefore, the intermediate stages in the development of the universe are not mentioned in Scripture.

The Development of the Sun

The bright sun that we see today looks very different from the early sun that was first formed. In particular, the present-day sun is much brighter and hotter than the early sun. The sun generates heat and light through a process called "nuclear fusion." In this process, atoms of hydrogen are combined together ("fused") to form atoms of helium. Over long periods of time, as the sun fused more and more hydrogen to form helium, the sun gradually became hotter and brighter. The sun that we observe today required billions of years to develop.

The point is that the early cooler and dimmer sun was sufficiently bright to produce enough sunshine to permit the process of photosynthesis to take place in plants. Therefore, plants could flourish before the development of the bright, present-day sun.

Another Question

There remains another question. How did the earth remain warm enough for animal life to exist if the sun was originally so much cooler? The answer is that the earth was covered with a layer of water. But why did this water remain liquid, rather than freezing into ice, because the sun was so cool? It was the atmosphere that protected the surface water from freezing. The atmosphere contains "greenhouse gases" that keep the surface of the earth warm. These gases acted as a blanket over the earth that preserved its heat and prevented the surface water from freezing. Today, "greenhouse gases" are a threat to our climate, because they cause "global warming." But long ago, the "global warming" caused by "greenhouse gases" was precisely what was needed to keep the earth warm and thus to permit animal life to exist on our planet.

The Sun in Scripture

According to the principle mentioned earlier, Scripture does not deal with intermediate stages in the development of the sun and *mentions only the sun that one observes today*. The dazzlingly bright sun that one observes today reached its present form only after the appearance of the earliest plants. Therefore, the sequence of events given in Scripture — first the plants and then the present-day sun — corresponds correctly to the order of events that occurred in the development of the solar system.

This same principle answers other questions as well. Scripture mentions only one moon — ours. But it is now known that in the solar system, there are over one hundred moons that revolve around their mother planet. Why does Scripture not mention them? The explanation is that before the invention of the telescope in the seventeenth century, these other moons could not be seen from the earth and no one knew of their existence. Scripture mentions only what could be observed by the inhabitants of the ancient world

For the same reason, Scripture makes no mention of the other planets in our solar system. In olden times, it was thought that the other planets were stars. They seemed to be very strange stars because they moved between the constellations, whereas all other stars always remained fixed in the same position relative to each other and relative to their constellation. Therefore, the ancients referred to most stars as "fixed stars," whereas the planets were called "wandering stars." The word "planet" is Greek for "wanderer." Even today, the terms "evening star" and "morning star" are both used to refer to the planet Venus, which is, of course, not a star at all. Thus, ancient astronomers thought that the only heavenly bodies were the sun, the moon, and the stars. And these are the heavenly bodies that are mentioned in Scripture.[12]

---

[1] I. Ribas, February 2010, *Proceedings of the International Astronomical Union*, vol. 264, pp. 3–18.

[2] A. Pavlov *et al.*, May 2000, *Journal of Geological Research*, vol. 105, pp. 11981–11990.

Summary

We have shown that the order of events that appears in the Genesis creation narrative is consistent with scientific discoveries. We have dealt here with the Third Day of Creation and the Fourth Day of Creation.

Chapter 10

The Creation of the Universe for the Benefit of Mankind

Several decades ago, scientists noticed that the universe has many features that make it appear as if the universe had been specifically designed to permit the existence and the flourishing of mankind. This phenomenon, which has attracted considerable scientific attention, is known as the anthropic principle,[1] from the Greek word *anthropos*, meaning "man." There are three aspects to the anthropic principle:

- Very slight changes in the laws of nature would have made it impossible for life to exist.
- Very specific conditions are required for the existence of life.
- Human life became possible only through the occurrence of a large number of highly improbable events.

Whereas the secular scientist views these three occurrences as being mere lucky events, the believing person sees in them the guiding hand of God.

The discussion consists of two parts:

---

[1] J. D. Barrow and F. J. Tipler, 1986, *The Anthropic Cosmological Principle* (Oxford University Press: Oxford); G. Gale, December 1981, *Scientific American*, pp. 114–122.

- A scientific explanation of exactly what is meant by the anthropic principle.
- The importance of the anthropic principle for the believing person.

The first topic is purely scientific, whereas the second topic deals with religion. This distinction must be kept clear because ironically, the words used by secular scientists in discussing the anthropic principle sound remarkably similar to words used by clergy!

The Laws of Nature and the Existence of Life

The anthropic principle includes a deep connection between the laws of nature and the existence of life. A relationship between the principles of biology and the existence of life is not surprising, but one would not expect to find such a connection when it comes to the physical sciences. But we shall see that the existence of life is intimately dependent on the details of the laws of physics, astronomy, and cosmology.

Why Does the Sun Shine?

It is not necessary to elaborate on the fact that life on Earth is crucially dependent on the sun, whose heat and light are the primary source of all terrestrial energy (aside from radioactivity, which is not relevant to our discussion). Without solar energy, our planet would be incapable of supporting life. Therefore, we begin our discussion of the anthropic principle by examining the mechanism that produces the sun's energy.

It should be mentioned that the sun is an ordinary star. Therefore, what will be stated about the sun applies equally well to every star in the universe.

The sun contains only two kinds of atoms: hydrogen and helium.[2] Helium is inert and, therefore, need not concern us further.

---

[2]Although I use the term "atom," the extreme temperature of the sun strips the electrons from the atom, leaving only the nucleus.

Our discussion centers on hydrogen, the simplest atom, whose nucleus consists of a single proton. The sun is basically an enormous ball of protons. How protons produce solar energy was explained by Hans Bethe, who was awarded the Nobel Prize for his discovery. (Bethe was a German Jew, and like all other Jews, he was dismissed from his university professorship by the Nazis. He then settled in the United States and joined the physics faculty of Cornell University, where he made his Nobel Prize-winning discovery.)

Because of the extreme conditions in the interior of the sun, a proton occasionally transforms into a neutron, another fundamental particle of nature. The resulting neutron can combine with another proton to form a composite particle known as a deuteron. Deuterons "burn" via a thermonuclear reaction, and this burning provides the intense heat and brilliant light of the sun. Deuterons are the solar "fuel" that generates the energy of the sun which enables life to exist on Earth.

An important feature of the thermonuclear burning of deuterons is that it occurs *very gradually*, thus ensuring that the sun generates energy slowly and will continue to shine for a very long time.

Another nuclear reaction that could, in principle, occur is the combination of one proton with another proton. Fortunately for us, proton–proton combination does not occur. If protons were able to combine with each other, then all the protons in the sun would immediately combine, leading to a gigantic explosion of the entire sun, and the destruction of the entire solar system, including the planet Earth.

The possibility of a proton–neutron combination and the impossibility of a proton–proton combination both depend on the strength of the *strong nuclear force*, which is the force between protons and neutrons. The other forces of nature are the electromagnetic force, gravity, and the weak nuclear force. Detailed calculations of the strong nuclear force lead to the following results.[3]

- If the strong nuclear force were only a *few percent* weaker, then a proton could not combine with a neutron to form a deuteron.

---

[3] P. Davis, 1972, *Journal of Physics*, vol. 5, pp. 1296–1304.

If this were the case, no deuterons would be formed in the sun, and there would be no solar fuel, and the sun would not shine. Rather, the sun would merely be a cold ball of inert protons, thus precluding the possibility of life.

- If the strong nuclear force were only a *few percent* stronger, then all the sun's protons would rapidly combine with each other. If this were to occur, the sun would explode, once again precluding the possibility of life on Earth. Since the sun is a typical star, and all stars shine through the same mechanism of thermonuclear burning of their protons, there could not be life anywhere in the universe.

- The strength of the strong nuclear force *just happens* to lie within the narrow range in which neither of these two catastrophes occurs. This force is strong enough to cause protons to combine with neutrons, thus permitting thermonuclear burning which makes all stars shine, including the sun. However, this force in not strong enough to permit protons to combine with each other, which would cause every star, including the sun, to explode.

In summary, the proton–proton explosion of the sun *does not* occur, whereas the gradual thermonuclear burning of deuterons *does occur* in the sun, thus providing the warmth and light that are vital for life to exist on Earth. This is our first example of the anthropic principle.

Water and Air on Our Planet

It is not necessary to elaborate on the necessity of water and air for the existence of life. The Earth is blessed with an abundant supply of both, permitting life to flourish here, whereas our two neighboring planets in the solar system, Venus and Mars, are both devoid of water and air, and hence devoid of life. These facts may not seem particularly noteworthy, but we shall show just how remarkable they really are.

The space program has taught us many things about our solar system. One of the most interesting discoveries is that shortly after they were formed, Earth, Venus, and Mars all had large amounts of surface water. The deep channels observed today on the surface of Mars were carved out long ago by copious fast-flowing Martian surface waters.[4] Also, Venus was once covered by deep oceans that contained the equivalent of a layer of water three kilometers deep over its entire surface.[5] However, in the course of time, all surface water on Mars and Venus disappeared. How did our planet escape this catastrophe?

The surprising answer is that the Earth escaped this catastrophe by sheer chance! Earth *just happens* to be sufficiently distant from the sun that our surface water did not evaporate, as happened on Venus. Moreover, Earth *just happens* to be sufficiently close to the sun that the temperature remains high enough to prevent the oceans from freezing permanently, as happened on Mars. Therefore, Earth *alone*, among the planets of the solar system, is capable of supporting life.

Let us now consider the atmosphere. The planetary atmosphere is controlled by a very delicate balance, involving the subtle interplay of many factors.[6] This balance is so delicate that if the Earth were only slightly closer to the sun, its surface temperature would be higher than the boiling point of water, ejecting the gases of the atmosphere into outer space, thus precluding all possibility of life. Moreover, if the Earth were only slightly farther from the sun, the concentration of carbon dioxide in the atmosphere would become so high that "the atmosphere would not be breathable by human beings."[7] Fortunately, the Earth's orbit *just happens* to lie at the crucial distance from the sun that permits the formation of a

---

[4]J. Audouze *et al.* (eds.), 1985, *The Cambridge Atlas of Astronomy* (Cambridge University Press: Cambridge), pp. 124–149.

[5]J. Audouze *et al.* (eds.), 1985, *The Cambridge Atlas of Astronomy* (Cambridge University Press: Cambridge), pp. 70–81.

[6]J. F. Kasting *et al.*, February 1988, *Scientific American*, pp. 46–53.

[7]J. F. Kasting *et al.*, February 1988, *Scientific American*, p. 53.

life-sustaining atmosphere ("life could appear in this extremely narrow zone"[8]).

These remarkably fortunate coincidences are known to scientists as the "Goldilocks problem of climatology." Recall the children's story in which Goldilocks entered the home of the three bears and found that the various items of Baby Bear were "not too hot and not too cold, not too hard and not too soft, not too long and not too short, not too high and not too low, but *just right.*" In the same vein, scientists view the existence of water and air on Earth as another example of the anthropic principle.

Physics and Astronomy

Many more examples of the anthropic principle could be brought from the physical sciences. Indeed, the examples are so numerous that many scientists have commented on the severe restraints that the laws of nature place on the existence of life. Particularly perceptive are the impressions of Freeman Dyson, of the Institute for Advanced Study in Princeton (where Albert Einstein was a professor for many years), whose words capture the very essence of the anthropic principle[9]:

> "As we look out into the universe and identify the many features of physics and astronomy that have worked together for our benefit, it almost seems as if the universe must have known that we were coming."

The Origin of Life

The branch of science that deals with the origin of life is called molecular biology, an area in which there has been enormous progress. Scientists have unraveled the structure of DNA (the long, thread-like molecules that form the genetic material found in each cell of every living creature) — the famous double helix. The genetic

---

[8]J. Audouze *et al.* (eds.), 1985, *The Cambridge Atlas of Astronomy* (Cambridge University Press: Cambridge), p. 63.

[9]F. Dyson, September 1971, *Scientific American*, p. 59.

code has been deciphered, explaining how genes control the production of proteins, which are the building blocks of the body. Hundreds of complex biochemical reactions that occur within the cell are now understood. With all this progress, one might easily conclude that the "riddle of life" has been solved. In other words, it might be concluded that scientists have now succeeded in explaining the steps by which inanimate material became transformed into the complex biological system that we call life. However, such a conclusion would be erroneous.

Many decades of intensive research in molecular biology has led scientists to appreciate how difficult it is to explain how inanimate material could have been transformed into living cells. This was the theme of a *Scientific American* article, appropriately entitled, "In the Beginning." The article describes the enormous difficulties encountered by all scientific proposals for the origin of life, quoting leading experts.

Harold Klein, chairman of the U.S. National Academy of Sciences committee that reviewed origin-of-life research, is quoted as follows[10]:

> *"The simplest bacterium is so complicated that it is almost impossible to imagine how it happened."*

Francis Crick, who shared the Nobel Prize for discovering the structure of DNA, is also quoted[11]:

> *"The origin of life appears to be almost a miracle, so many are the conditions which have to be satisfied to get life going."*

If this Nobel laureate, a man completely devoid of any religious feeling, uses the words "*almost a miracle*" to describe the origin of life, it is clear that an incredible series of highly unlikely conditions must have been satisfied to produce living cells.

---

[10] H. P. Klein, February 1991, *Scientific American*, p. 104.
[11] F. Crick, February 1991, *Scientific American*, p. 109.

This view of Crick is shared by many other biologists. For example, a scientific article states[12]:

> *"How single-celled organisms formed from their bare chemical ingredients remains one of the great mysteries of biology."*

Highly Improbable Events and Human Beings

The Destruction of the Dinosaurs

Thus far, we have been discussing the laws of nature and the many unlikely events that were necessary for the existence of life. But our concern is, of course, with human life. Did any highly improbable events occur that permitted the existence of human beings? We shall see that scientists answer with a resounding "Yes!" This is the heart of the anthropic principle.

We begin with a discussion of the dinosaurs, those terrible monsters of the past. The dinosaurs were one of the most successful groups of large animals that ever lived — the largest, strongest, and fiercest animals of their time. Dinosaurs (and their close relatives) inhabited every continent, the air (flying dinosaurs), and the oceans (marine dinosaurs). All other animals lived in constant fear of being destroyed by these gigantic reptiles. Because dinosaurs dominated the Earth, this era is called the Age of Reptiles.

After being the undisputed masters of our planet for 150 million years, the dinosaurs suddenly disappear from the fossil record. The destruction of all the world's dinosaurs, together with most other animal species, is the most famous of the mass extinctions that have occurred periodically in the history of our planet. A mass extinction is an event that wipes out the majority of all animal species.

The cause of the mass extinction that destroyed all the dinosaurs had baffled scientists for many years. What worldwide catastrophe caused the abrupt demise of these extremely successful animals that had enjoyed such a long period of dominance?

---

[12] March/April 2005, *Europhysics News*, p. 17.

After years of debate, the riddle of the sudden destruction of all the dinosaurs was solved in 1980 by Nobel laureate Luis Alvarez and his son Walter. These two scientists showed that a giant meteor from outer space collided with Earth to cause this worldwide catastrophe.[13] This explanation for the mass extinction — the impact of a meteor colliding with the Earth — is known as the impact theory. Evidence in favor of the impact theory accumulated rapidly, and Luis Alvarez could soon point to *fifteen* different pieces of scientific data that support this theory.[14]

The point of central importance to our discussion is that the collision of the meteor with Earth appears to be a matter of *sheer luck*. This has been emphasized by leading paleontologists (scientists who study fossils). David Raup, past president of the American Paleontological Society, made this point the central theme of his famous article (later expanded into a book), entitled *Extinction: Bad Genes or Bad Luck?* In his book, Raup stresses the importance of luck in mass extinctions, writing[15]: "*As is surely clear from this book, I feel that most species die out because they are unlucky.*"

The important role played by luck in mass extinctions has also been emphasized by Stephen Jay Gould of Harvard University[16]:

"If extinctions can demolish more than 90% of all species, then we must be losing groups forever by pure bad luck."

George Yule, of the University of Oxford, put it in the following way[17]:

"The species exterminated were not killed because of any inherent defects, but simply because they had the ill-luck to stand in the way of the cataclysm."

---

[13] W. Alvarez *et al.*, 1984, *Science*, vol. 223, pp. 1135–1140.

[14] L. Alvarez, July 1987, *Physics Today*, pp. 24–33.

[15] D. Raup, 1991, *Extinction: Bad Genes or Bad Luck?* (Oxford University Press: Oxford), p. 191.

[16] S. J. Gould, 1981, *The Flamingo's Smile* (W. W. Norton: New York), p. 242.

[17] G. Yule, 1981, *Philosophical Transactions of the Royal Society*, vol. 213, p. 24.

Finally, we quote David Jablonski, of the University of Chicago, renowned authority on the subject of mass extinctions[18]:

> *"When a mass extinction strikes, it is not the "most fit" species that survive. Rather, it is the most fortunate. Species that had been barely hanging on, suddenly inherit the Earth."*

These leading scientists all emphasize that if a giant meteor suddenly falls from the sky, wiping out some species while permitting other species to survive and ultimately to flourish, then the Darwinian principle of "survival of the fittest" is irrelevant. The species that survived were simply blessed with *good luck* — the occurrence of an extremely improbable and totally unexpected event that destroyed their competitors.

The Dinosaurs and Man

What is the relationship between the dinosaurs and human beings? Why does the sudden destruction of all the dinosaurs worldwide constitute a dramatic example of the anthropic principle? The explanation is straightforward. As long as dinosaurs dominated the Earth, there was no possibility for large mammals, including humans, to exist. Only after the dinosaurs were wiped out could the mammals flourish and become the dominant fauna.

This intimate connection between human beings and dinosaurs was emphasized by Nobel laureate Luis Alvarez, who ended his article about the meteoric impact that destroyed all the dinosaurs, with the following words[19]:

> *"From our human point of view, that impact (that destroyed the dinosaurs) was one of the most important events in the history of our planet. Had it not taken place, the largest mammals alive today might still resemble the rat-like creatures that were then scurrying around trying to avoid being devoured by dinosaurs."*

---

[18] D. Jablonski, June 1989, *National Geographic*, p. 673.
[19] L. Alvarez, July 1987, *Physics Today*, p. 33.

But there is more to the story. For human beings to exist today, it was not sufficient that our planet had been struck by a large meteor. The impact had to have *just the right force*. Alvarez writes[20]:

> *"If the impact had been weaker, no species would have become extinct. Mammals would still be subordinate to the dinosaurs, and I (Alvarez) wouldn't be writing this article. If the impact had been stronger, all life on this planet would have ceased, and again, I wouldn't be writing this article. However, the impact had just the right strength to ensure that the mammals survived, while the dinosaurs didn't."*

Stephen Jay Gould

It has become clear to scientists that the sudden destruction of the world's dinosaurs was just one of a long series of completely unexpected, highly improbable events that had to occur for human beings to exist — and that *all these events just happened to occur in precisely the required sequence*. Stephen Jay Gould, of Harvard University, wrote a book on this topic, which abounds with examples of the anthropic principle. Gould repeatedly emphasizes how amazing it is that human beings exist at all, writing[21-24]:

> *"We are an improbable and fragile entity ... the result of a staggeringly improbable series of events, utterly unpredictable and quite unrepeatable."*
>
> *"Consciousness would not have appeared on our planet if a meteor had not claimed the dinosaurs as victims. We owe our existence, as large reasoning mammals, to our luck."*
>
> *"Let the 'tape of life' play again from the identical starting point, and the chance is vanishingly small that anything like human intelligence would grace the replay."*

---

[20] L. Alvarez, July 1987, *Physics Today*, p. 29.
[21] S. J. Gould, 1989, *Wonderful Life* (W. W. Norton: New York), pp. 14, 319.
[22] S. J. Gould, 1989, *Wonderful Life* (W. W. Norton: New York), p. 318.
[23] S. J. Gould, 1989, *Wonderful Life* (W. W. Norton: New York), p. 14.
[24] S. J. Gould, 1989, *Wonderful Life* (W. W. Norton: New York), p. 289.

"It fills us with amazement that human beings exist at all. Replay the tape a million times from the same beginning, and I doubt that Homo sapiens *would ever appear again."*

Dissenting Voices and Their Errors

Although the anthropic principle is widely accepted, dissenting voices have been heard. The criticism most commonly raised against the anthropic principle displays a misunderstanding of probability theory, which is known for being notoriously subtle. Examples follow.

A professor of philosophy from the University of North Carolina wrote[25]:

"I pull a $1 bill from my wallet and observe its serial number to be G65538608D. The probability for this occurrence is less than one in a billion. Thus, undeniably, I am faced here with an extremely rare event."

A professor of genetics from the Hebrew University in Jerusalem wrote[26]:

"Religious scientists place great emphasis on the remarkable coincidences which characterize the universe. The point of this claim is that such remarkable events could not have occurred through chance, but rather are the result of a guiding hand. Superficially, this claim appears convincing, but a little thought shows that that it is without foundation. According to this logic, the probability that I am writing these lines with a dull yellow pencil, using my left hand, sitting at my kitchen table, on the third floor of a specific Jerusalem address — this probability is completely negligible. Nevertheless, all these events happened and they clearly mean nothing."

Defining the Event — Richard Feynman

The conceptual error in the above criticisms has been clarified by Nobel laureate Richard Feynman, one of the most brilliant

---

[25] G. Schlesinger, Spring 1988, *Tradition*, vol. 23, p. 3.
[26] R. Falk, Spring 1994, *Alpai'im* (Hebrew), vol. 9, pp. 136.

physicists of the twentieth century. In his book on quantum electro-dynamics, a theory based on probabilities, Feynman writes[27]:

"In order to calculate correctly the probability of an event, one must be very careful to define the event clearly"

Following Feynman's sage advice, we shall define the event clearly. This leads to the conclusion that *the probability is 100%* that the dollar bill pulled from the wallet of the professor of philosophy has serial number G65538608D! Why? Because this number was chosen *after* looking at the serial number. Thus, the question that was really asked by the philosopher is the following: What is the probability that the serial number that I see on the dollar bill really *is* the serial number on the dollar bill? The answer is clearly: 100%.

The quote from the geneticist is wrong for the same reason. *The probability is 100%* that his article was written on the kitchen table, using a dull yellow pencil held in the left hand, on the third floor of a specific address. Why? Because these conditions were chosen *knowing* what had already happened. Thus, the question that the geneticist really asked is the following: "What is the probability that what I know to have happened, really did happen?" The answer to this question is clearly: 100%.

If one could correctly guess the serial number *before* one looks at the dollar bill, we would all be absolutely astonished — and with good reason! Similarly, if one could guess correctly all the conditions under which *someone else* had written an article, then we would all be flabbergasted — and rightly so.

Feynman humorously dismisses all claims similar to the above quotes from the professors of philosophy and of genetics in the following amusing manner:

"A most remarkable thing happened to me this morning. I noticed that the car parked next to mine had the following license number: FY56298. There are about 100 million cars in the United States, and only one car has

---

[27] R. Feynman, 1985, *QED — The Strange Theory of Light and Matter* (Princeton University Press: Princeton), p. 81.

> license number *FY56298*, and I happened to park next to that car! The
> chance of that event happening is 100 million to one — and it happened!
> Isn't that remarkable!"

Feynman then explains the error in this argument. What happened was that the car next to Feynman's had *some* license number.
But *every* car has some license number. Therefore, the chance is
100% that the car next to Feynman's had a license number.

Feynman's explanation also applies to the two quotes given
above because *every* article is written under *some* set of conditions
and *every* dollar bill has *some* serial number. Therefore, the chance
for the event occurring is 100%, and not "*less than one in a billion*" or
"*completely negligible*," as claimed by the two professors.

Further Invalid Criticism

A professor of probability, formerly of the Soviet Union and more
recently at California State University, wrote the following criticism
of the anthropic principle.[28] He recognized the error in the above
two quotes, but claimed that the same error is being made by the
proponents of the anthropic principle. His claim is that since we
already know that life exists, the chance of life having come into
existence must be 100%.

It is clear that Feynman's point has not been understood. What
is crucial to Feynman's explanation is that one is dealing with an
event that *always happens*. *Every* car has some license number, *every*
article is written under some set of conditions, and *every* dollar bill
has some serial number. Therefore, what happened was that one
result occurred out of a great many possible *equivalent* results (this is
known in statistics as "equivalent microstates"). That always happens
and there is nothing unusual about it.

Now we turn to the origin of life. In complete contrast to the
previous examples, *life did not have to come into existence at all.*
Therefore, the particular arrangement of molecules in the cell that

---

[28] M. Perakh, 2004, *Unintelligent Design* (Prometheus Books: Amherst, NY), p. 268.

leads to life is a *unique* microstate. The existence of life and the absence of life are *not* equivalent microstates.[29] Indeed, life requires an arrangement of molecules that is so unusual that Nobel laureate Francis Crick called its occurrence, "almost a miracle."

Events in Context — Playing the Lottery

We now discuss the second important aspect of Feynman's statement: *events must be defined in context.* An example will illustrate this point.

Consider a lottery in which one million people buy a lottery ticket each week, and one lucky ticket holder wins the grand prize. If George wins this week's lottery, nobody will be surprised. But why not? The probability that George would win the lottery was only one in a million — and it happened! The reason for the lack of surprise is that each of the one million lottery players is completely equivalent. Although the probability was only one in a million that the winner would be George, there are one million "equivalent" Georges ("equivalent microstates"). Therefore, what really happened was that *someone* won the lottery. The probability for that event happening — *someone* winning — is 100%. Hence, there was no reason to be surprised.

If George would *again* win the lottery in the following week, everyone would most certainly be amazed because the *context* is entirely different. In the first week, George was just one of a million equivalent lottery players. But in the second week, he was a *unique* individual — the fellow who won last week — and the probability of this unique individual winning the lottery again is truly one in a million.

If George were to *again* win the lottery for the *third* consecutive week, it is clear that *suspicion*, not surprise, would be the natural reaction. The probability that the same person will win the lottery

---

[29] In principle, one could argue that life is not a unique phenomenon and there is no significant difference between living creatures and inanimate objects. Cats and chairs, it's all the same. However, I doubt that many people hold such a view.

three times in a row is so extremely small that one suspects that a guiding hand was behind George's triple win.

The Importance of George's Triple Win

The aspect of George's triple win that arouses suspicion is *not* that a very rare event has occurred. Extremely rare events occur all the time. It is the *repeated* occurrence of rare events that arouses suspicion. George won the lottery *again and again.*

Scientists who discuss the anthropic principle always call attention to the *many* unusual features of the universe or the *many* events that are necessary for life in general and for human life in particular, and all these *many* required features/events occurred. Freeman Dyson: "the *many* peculiarities of physics...." Francis Crick: "so *many* are the conditions...." Stephen Jay Gould: "a staggeringly improbable *series* of events...."

The Anthropic Principle and the Believing Person

What are the implications of the anthropic principle for the believing person? Scientists have discovered that the existence of human beings requires many severe constraints on the laws of nature and it also required the occurrence of many extremely unlikely events: "a staggeringly improbable series of events," wrote Stephen Jay Gould. The extreme rarity of events required for man's existence is well established. That is the content of the anthropic principle. But before deciding on the *significance* of these events, one must first understand the *significance* of human beings.

If human beings are assumed to be no more important than any other of the two million animal species identified so far, then the anthropic principle would have no meaning. Rarity by itself is not significant. However, Scripture views mankind very differently from all other species, and emphasizes that human beings were the ultimate purpose of the Creation (Genesis 1:28):

"God blessed Man and said: 'Be fruitful and multiply. Fill the land and conquer it. Rule over the fish of the sea and over the winged creatures of the sky, and over every living creature that crawls upon the earth.' "

If one accepts the centrality of human beings, then the anthropic principle is of the utmost significance and constitutes yet another example of the harmony between Scripture and modern science.

Chapter 11

Problematic Numbers in Scripture I: The Extreme Longevity of the Early Generations in Genesis

Questions

One of the most difficult questions in Scripture relates to the extreme ages ascribed to the twenty generations from Adam to Abraham. Genesis speaks of people in this period living for more than 900 years, culminating in the record holder, Methusaleh, who reached the unbelievable age of 969 years. How can one understand such longevity? Anyone who has had close contact with the very elderly has observed that the functioning of the human body deteriorates when approaching the age of 100. Thus, the account in Scripture of people living for many hundreds of years seems completely impossible.

There is yet another difficulty. After the Exodus from Egypt, extreme longevity disappears and the life span of subsequent generations becomes normal by present-day standards — completely consistent with the traditional 120-year maximum life span. What happened to cause this dramatic decrease in longevity?

These are the questions to be addressed in this chapter. It will be shown that recent scientific advances regarding the process of ageing pave the way to understanding the thousand-year life spans of the early generations in Genesis, as well as the decrease to contemporary life spans of those who lived after the time of the Exodus.

Living Longer

Why do human beings age? Until quite recently, no one really knew. However, in the last few decades, there have been enormous advances in our understanding. For a recent (2018) survey of research in the study of ageing, see *Conn's Handbook of Models for Human Ageing.* The study of ageing has become the focus of such intense scientific effort that a leading authority speaks of "a revolution in ageing research."[1] Some of the findings of this research have been so completely unexpected that scientific journals aimed at the educated layman audience now abound with articles that describe these exciting discoveries. Some examples will illustrate the point.

From the cover of *New Scientist: "Life at 200: Will We Always Grow Old?"* The cover story, dramatically entitled *Death of Old Age,* begins as follows: *"We can live healthy lives well into our hundreds, researchers claim."*[2]

A news item in *Scientific American,* entitled *Immortality Gene Revealed,* states: *"Two teams of scientists have cloned the gene for telomerase, known as the 'holy grail' of ageing research … Cells that produce telomerase are immortal."*[3]

Michal Jazwinski, Director of the Center on Ageing at Louisiana State University, a major figure in ageing research, asserts that *"the maximum human life span might go as high as 400 years."*[4]

Why Do We Age?

Ageing is a universal human experience. Ageing and death seem as natural as breathing, and just as inevitable. Although ageing was long regarded as a mysterious aspect of life, scientific research has now revolutionized our understanding about what causes it.[5]

---

[1] S. M. Jazwanski, 1996, *Science,* vol. 273, p. 54.

[2] D. Consar, 22 June 1996, *New Scientist,* p. 24.

[3] "News in Brief," October 1977, *Scientific American,* p. 14.

[4] R. L. Rusting, December 1992, *Scientific American,* p. 95.

[5] For a popular account of the many recent advances in ageing, see R. E. Ricklefs and C. Finch, 1995, *Ageing* (Scientific American Library: New York).

Scientists continue to make breathtaking progress and gain new insights about the basic mechanisms responsible for ageing. More importantly, they now have the ability to intervene in the ageing process and thereby extend the human life span. This newfound knowledge strongly suggests that biomedical advances will eventually enable us to delay and even eliminate altogether many of the causes of ageing and death.

The characteristics of ageing are many. The body produces chemicals (free radicals) that destroy tissues by a process called oxidation. The immune system weakens and is no longer able to defend the body against disease. Structural proteins become altered, leading to rigidity of the heart muscles, lungs, ligaments, and tendons. Certain cells lose their ability to divide (Hayflick limit). DNA molecules, which are vital for cellular replication, become damaged by mutations. Cancers develop as cells suddenly proliferate out of control. Hormonal changes occur that cause the gradual destruction of bones (osteoporosis). Critical enzymes cease functioning. Strokes attack the brain. Arthritis appears in the joints. Nerve cells in the brain degenerate (Alzheimer's disease). Blood vessels lose their elasticity (arteriosclerosis) and cease to function properly. Parkinson's disease and diabetes develop. Memory declines. And the list goes on.

It now appears that there is a common cause for this seemingly endless list of afflictions of old age. A scientific consensus is emerging that the root cause of all ageing processes is genetic. According to Caleb Finch, of the Department of Neurobiology of Ageing at the University of Southern California[6]:

"We are convinced that the rate of ageing is under genetic control."

The body *does not wear out* in the way that a car or a washing machine wears after years of faithful service. Rather, the human body contains certain genes that cause all the havoc of old age listed above.

---

[6]R. E. Ricklefs and C. Finch, 1995, *Ageing* (Scientific American Library: New York), p. 176.

In other words, we all suffer from genetic defects. If our defective genes could be identified, and their effects neutralized through genetic engineering, the human life span could be extended, perhaps even significantly. This exciting possibility, discovered by scientists, is responsible for the dramatic pronouncements quoted earlier.

The idea that genes cause ageing has received important support from the research of Mark Azbel, formerly at Moscow State University and now at Tel Aviv University. In a series of pioneering articles, Azbel showed that the extensive mortality data for human beings can all be explained by assuming a genetic basis for ageing and death, writing[7]:

> *"There exists a genetically programmed probability to die at a given age ... This age may be genetically manipulated."*

Experiments that alter the genetic structure of laboratory specimens have already begun, producing striking results. A favorite subject for study is a small nematode worm (*Caenorhabditis elegans*) that has 13,000 genes. Tom Johnson, of the University of Colorado, found that altering a single gene, aptly named *age-1, doubles the life span* of this nematode.[8]

Michael Rose, of the University of California, has genetically engineered a new strain of fruit flies (*Drosophila melanogaster*) that *live almost twice as long* as standard laboratory-reared flies. Moreover, Rose found that[9]:

> *"These 'superior' flies are more robust at every age. Even when old, many of these flies are stronger than ordinary young specimens."*

Similarly, Michal Jazwinski has identified several genes that prolong the life of brewer's yeast (*Saccaromyces cerevisiae*). Introducing the gene *LAG-1 significantly extends the life span,* and he also found that[10]:

---

[7]M. Ya. Azbel, 1997, *Physics Reports*, vol. 288, p. 545.
[8]T. R. Johnson, 1988, *Genetics*, vol. 118, pp. 75–86.
[9]R. L. Rusting, December 1992, *Scientific American*, p. 87.
[10]R. L. Rusting, December 1992, *Scientific American*, p. 91.

"Yeast cells that bear this gene maintain their youth longer."

For the reader who finds incredible the suggestion that genetically engineered living creatures could have enormous life spans, we point out that *even now*, there are many animals that do not exhibit any signs of ageing, and they continue to bear offspring for as long as they live. As Leonard Hayflick, of the University of California, points out[11]:

> *"Some animals do not seem to age at all. If they do age, it occurs at such a slow rate that ageing has not been demonstrated. These non-ageing animals experience a peak in their physiological functions, but these functions then do not seem to decline ... However, non-aging animals do not live forever because of accidents, disease and predation."*

Perhaps the most astonishing data come from the field studies of the Scottish ornithologist George Dennet, who has spent a lifetime observing a colony of marine birds called fulmars (*Fulmaris glacialis*) on the Orkney Islands. Dennet reports that[12]:

> *"Fulmars show no increase in mortality rate and no decline in reproduction up to at least 40 years. Certainly no species of similarly sized mammals or birds maintain their fertility at a comparable age. Do these birds avoid ageing altogether? We do not know."*

Studies of tortoises and certain fish yielded similar results[13]:

> *"One specimen of Marion's tortoise (Geochelone gigantea) died accidentally (run over by a truck) at age 150 in a British military fort on Mauritius. Studies in progress on other tortoise species suggest that they remain fertile throughout their long life and that their mortality rate remains low."*

---

[11] L. Hayflick, 1994, *How and Why We Age* (Ballantine Books: New York), p. 22.

[12] G. M. Dunnet, 1988, in *Reproductive Success: Studies of Individual Variations in Contrasting Breeding Systems*, editor T. H. Clutton-Brock (University of Chicago Press), p. 268.

[13] R. E. Ricklefs and C. Finch, 1995, *Ageing* (Scientific American Library: New York), p. 8.

The longevity record for fish is held by the sturgeon (*Acipenser fulvescens*) which reaches the age of 150 years, as confirmed by the number of rings on their scales. The very old individuals of rock fish (*Sebates aleutianus*) studied by Bruce Leaman of the Pacific Biological Station of Fisheries produced egg masses and showed no signs of the tumors and other pathological lesions usually found in animals of advanced ages.[14]

A World Without Ageing

In the light of the facts presented above, it should not be too difficult to imagine a world in which human beings do not age. This does not mean that no one will ever die. Lives would still be cut short by the usual hazards of traffic accidents, virulent diseases, and violent crime. But the *rate* at which people die would not increase with age. For example, the chance of dying in a car accident are the about same at age 60 and at age 20.

The safest age for human beings is between ten and fifteen years. The period of childhood diseases is past and the infirmities of old age have not yet begun. One speaks of the mortality rate, which is defined as the chance of dying within the next year. In the United States and Western Europe, the mortality rate between the ages of ten and fifteen is about 0.05%. That is, only 1 in 2000 youngsters will die within a year. This is the minimum mortality rate observed for human beings. By contrast, the mortality rate for hundred-year-olds is 50%. Half the centenarians will not survive the year.

What can one say about the human life span in a world without ageing? If people did not age, then everyone would remain forever young, and the minimum mortality rate of a ten-year-old child would persist throughout one's entire life. Caleb Finch has shown that under these circumstances the average life span would be about 1200 years.[15] Moreover, chronologically extremely old, but biologically still young, men and women would be able to sire children

---

[14]R. E. Ricklefs and C. Finch, 1995, *Ageing* (Scientific American Library: New York), p. 10.

[15]R. E. Ricklefs and C. Finch, 1995, *Ageing* (Scientific American Library: New York), p. 2.

throughout their thousand-year lives. This is how society would be if one could eliminate all the genetic defects that cause ageing.

Scripture and Ageing

Having presented the scientific advances regarding ageing, we return to the account in Scripture regarding the extreme longevity of the early generations in Genesis. The life spans for the first twenty-six generations are given in the accompanying figure. It is clear that there is a marked difference between the life spans before Noah and after Noah. Up to and including Noah, the life spans are nearly the same (about 900–950 years), except for Enoch. However, Enoch is explicitly described in Genesis as having died young and may therefore be removed from consideration. After Noah, however, the life span decreases steadily, falling from 959 years for Noah (tenth generation) to the "traditional" value of 120 years by the time of Moses (twenty-sixth generation).

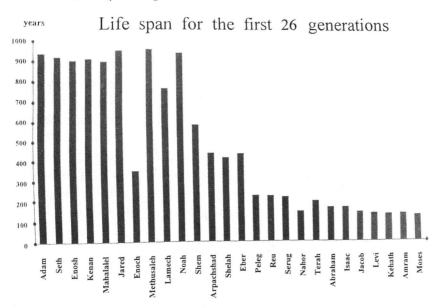

There are really two puzzles before us. The first is to explain the nearly 1000-year life spans until the generation of Noah. The second is to explain why these long life spans decreased steadily after the generation of Noah.

The Nearly Thousand-Year Life Spans Up to the Generation of Noah

Consider the following scenario. When Adam and Eve were in the Garden of Eden, they were destined to live forever. We propose that this immortality was due to Adam and Eve not possessing any of the genetic defects that nowadays cause ageing (discussed above). Moreover, Adam and Eve were not subject to the non-genetic causes of mortality that are unrelated to ageing. In the Garden of Eden, there presumably were no traffic accidents, no virulent diseases, and no violent crime. Therefore, it follows that Adam and Eve would live forever.

When Adam and Eve were banished from the Garden of Eden and forced to live in the "outside world," they became subject to the usual non-genetic causes of death that afflict us all (accidents, diseases, etc.) and thus they became mortal. Indeed, this is what God meant when He told Adam and Eve that "*on the day that you eat from the forbidden fruit, you will die*" (Genesis 2:17). Not that Adam and Eve would *die* on that day, but rather, that they would become *mortal* — subject to dying.

The key point of our proposal is that even after leaving the Garden of Eden, *Adam and Eve still did not have the genetic defects that cause ageing and, therefore, they did not age.* This explains both their great age longevity (930 years for Adam) and also their ability to bear children at a very advanced age (130 years for Adam).

We have already seen that in the absence of ageing, the average human life span is about 1200 years. The difference of only 30% between the contemporary non-ageing life span and the Scriptural life span should be viewed as good agreement.

This approach also explains the advanced age of childbearing listed for the early generations of Genesis. Methusaleh and Lamech were each nearly 200 years old when they sired children, and it is explicitly mentioned that they had additional children even later in life. Similarly, Noah was 500 years old when his three children were born. Moreover, at the age of 600 (Genesis 7:6), Noah was placed in charge of building and outfitting the Ark, hardly a task that could be assigned to a doddering old man. Thus, it is clear that Noah not only lived extremely long but also that he remained young throughout his life. As we have seen, unusually long life spans are consistent with the ability to sire children at an advanced age and the lack of signs of ageing.

The Steady Decrease in Life Span After the Generation of Noah

We now turn to our second challenge, namely, explaining the steady decrease in life span after the generation of Noah (see figure on page 105). As the figure shows, the life spans decreased after Noah (959 years), with no one living longer than 240 years after Eber (fourteenth generation). After Moses (twenty-sixth generation), there is no instance recorded in Scripture of anyone living significantly longer than 120 years, which remains to this day as the maximum human life span.

After the generation of Noah, men no longer sired children at an advanced age. After Shem (eleventh generation), the age of having children dropped to the thirties, just like today. Indeed, when Abraham and Sarah (twentieth generation) became parents at ages 100 and 90, respectively, Scripture describes the event as miraculous.

For the first time, people are described in Scripture as becoming old and infirm. Ageing in later life is a characteristic feature of the Patriarchs and Matriarchs (twentieth to twenty-second generations). This applied to Abraham and Sarah (Genesis 18:11–13), to Isaac (27:1–2), and to Jacob (48:10).

Clearly, something happened at the time of Noah that caused these important changes in later generations — changes that are all associated with ageing. Scripture reveals this event. Genesis 6:3 states that God was extremely displeased with the corrupt behavior of people at the time of Noah and, therefore, He decreed that the maximum human life span would be reduced to 120 years (*"Therefore, man's life span shall be for one hundred and twenty years"*).

In fact, the human life span did not immediately decrease to 120 years. A full ten generations after this divine pronouncement, Abraham still lived to age 175, with similar long life spans for the other Patriarchs.

In the light of our scientific discussion of ageing, we propose that the divine pronouncement of Genesis 6:3 can be understood as follows. At the time of Noah, the genes for ageing were introduced in the human gene pool. It would, of course, take a number of generations for these ageing genes to propagate throughout the entire human population. This explains why sixteen generations had to pass (from the generation of Noah until the generation of Moses) until the maximum life span finally became reduced to the divinely decreed value of 120 years. The human life span decreased steadily during this transition period, from 959 years for Noah to 120 years for Moses. This proposal also accounts for the fact that after the generation of Shem (Noah's son), there are no further (non-miraculous) instances in Genesis of people having children at an advanced age.

May You Live Until 120! A Blessing for Long Life

The traditional Jewish blessing for long life, the title of this section, is not merely an item of folklore. The source for this blessing is a verse in Scripture (Genesis 6:3) in which God decrees the maximum human life span to be 120 years. Since everyone presumably wants to live for as long as possible, it has become traditional for one to wish a friend the maximum number of years — in good health, of course. We shall examine the scientific basis for this number, asking whether there is any justification for assuming that 120 years is the maximum human life span.

It should be mentioned at the outset that the source for the traditional 120 years *cannot* be empirical. It is simply not true that the oldest people have always lived to about the age of 120. Before the twentieth century, anyone living for more than 100 years was an extreme rarity, and absolutely nobody reached the age of 110.

In fact, until quite recently, the scientific literature quoted the figure of either 105 years[16] or 110 years[17] for the maximum human life span. Therefore, on the basis of observation, there would never have been any reason to consider the higher figure of 120 years.

It should be emphasized that there is no factual basis whatsoever for the anecdotal reports of remote mountain villages of the Caucasus region of Georgia or in Ecuador or in Colombia or in Pakistan where people supposedly live much longer than 120 years. Such reports are very beneficial to the tourist trade, but investigation always reveals that these claims of extreme longevity are groundless.[18]

To promote the publicity value of such claims, both Russia and Colombia issued postage stamps in 1956 to honor their longest-lived citizen.[19] The Colombian stamp depicts a man said to be 167 years old, whereas the Russian stamp is somewhat more modest, claiming only the age of 148 for its most senior citizen. It is usually implied that the healthy atmosphere of the mountains, the relaxed lifestyle, and simple diet — sometimes a certain type of yogurt is mentioned — are responsible for these utterly fictitious ages!

The Maximum Human Life Span

It was only in the latter part of the eighteenth century that systematic birth records began to be kept on a widespread scale, thus making it possible to establish reliably the age of the oldest human beings. Based on such records, scientists had until recently accepted[20] that the oldest person in history was a Japanese man, Shigechiyo

---

[16]F. A. Lints, 1978, *Genetics and Ageing*, vol. 17 of *Interdisciplinary Topics in Gerontology*, series editor: H. P. van Hahn (S. Karger: Basel), p. 29.

[17]L. C. Campanelli, 1985, in *Aging*, (ed.) C. B. Lewis (Davis: Philadelphia), p. 7.

[18]L. Hayflick, 1994, *How and Why We Age* (Ballantine Books: New York), pp. 196–202.

[19]A. Lindenbaum, 1970, *Stamps That Tell a Story* (Sabra Books: New York), p. 15.

[20]D. W. E. Smith, 1989, *Biological Reviews*, vol. 64, p. 9.

Izumi, listed in the 1987 *Guinness Book of World Records*, who died in 1986 at age 120.

This record was broken[21] by a Frenchwoman, Jeanne Calment, born in 1875, who died in 1997 at the age of 122. For the first time in post-biblical history, we have authenticated cases of persons reaching the age of 120, confirming the divine pronouncement of Genesis 6:3.

Let us now consider the reverse question. Perhaps the figure of 120 years given in Scripture is, in fact, *too low* for the maximum human life span? With improved health care for the elderly and the medical miracles that we constantly witness, perhaps the day will come when people will live to the age of 130 or 140 or even beyond? Is there any scientific evidence in support of the biblical figure of *only* 120 years for the maximum human life span?

Making accurate estimates of the largest possible values of various quantities on the basis of previous experience is a new branch of statistics, known as *extreme value theory*. For example, it is important to be able to determine how destructive the next earthquake or flood will be, on the basis of the history of earthquakes and floods in the region. This field of statistics has received impetus through the research work of Richard Smith of the University of North Carolina.[22]

One of the most intriguing applications of extreme value theory is to the question of human longevity, a subject of great interest to both biologists and actuaries.[23] An analysis was recently carried out by Laurens de Haan and his colleagues at the Erasmus University in Rotterdam, Holland. Using all the available data on human longevity, de Haan found the value of 119±6 years for the maximum

---

[21] T. T. Perls, January 1995, *Scientific American*, p. 52.

[22] See, for example, R. Smith, 1990, *Extreme Value Theory*, in *Handbook of Applicable Mathematics: Supplement* (Wiley: New York).

[23] Extreme value theory determines the maximum value of a particular parameter by extrapolating known data for a given system, assuming that no new feature has been introduced. Thus, if genetic engineering were used to "improve" human beings, then the maximum life span calculated by extreme value theory would no longer be valid, since this calculated age refers to "unimproved" human beings.

human life span.[24] This result is consistent with value of 120 years given the Scripture.

The Oldest Old

Many have an unfavorable image of extremely old people. We think of nonagenarians and centenarians (often referred to as "the oldest old"[25]) as physically infirm and mentally debilitated. Thus, living to 120 years seems of doubtful value. Why should anyone want to live so long only to end up as a burden to his or her children?

Recent studies of the oldest old demonstrate that this unfavorable image has no basis in fact. Thomas Perls, a geriatrician at Harvard Medical School, explains[26]:

> *"The prevailing view of ageing as advancing infirmity is wrong ... The oldest old are often the most healthy and agile of the senior people in my care ... Centenarians, with few exceptions, report that their nineties were essentially problem-free. As nonagenarians, many were employed, sexually active, and enjoyed the outdoors and the arts. They carry on as if age were not an issue. Accumulating evidence indicates that we must revise the common view that advancing age inevitably leads to extreme deterioration."*

Similar views are expressed by Richard Suzman of the National Institute of Ageing at the NIH, who emphasizes that the pessimistic expectations of infirmity for the oldest old are being revised in the light of new studies.[27]

> *"National data show that a surprisingly large percentage of the oldest old not only require no personal assistance on a daily level, but are also physically robust ... Health-service research is revealing that a large percentage of those in old age remain low-cost users of medical services."*

---

[24] R. Matthews, 12 October 1996, *New Scientist*, p. 40.

[25] R. M. Suzman, D. P. Willis and K. G. Manton (eds.), 1992, *The Oldest Old* (Oxford University Press: Oxford).

[26] T. T. Perls, January 1995, *Scientific American*, p. 50.

[27] R. M. Suzman, 1992, in *The Oldest Old*, (eds.) R. M. Suzman, D. P. Willis and K. G. Manton (Oxford University Press: Oxford), p. 343.

Other authorities confirm these views[28]:

"New scientific evidence regarding physiological ageing shows that we had overestimated the rate of decline for the oldest old in various physiological functions. Studies of healthy persons of advanced ages show that physiological functions of many types decline much more slowly than previously thought. For example, the cardiovascular function of healthy 80-year-olds was found to be not much different from that of 30-year-olds."

I would like to add a personal testimony. In 1984, I had the great pleasure of meeting Charlotte Hughes, then Great Britain's third oldest person at the age of 111. Mrs. Hughes was physically fit and completely lucid throughout our conversation. She told me that she reads from Shakespeare and the Bible every day. On the occasion of her 110th birthday, she was flown to New York City on a supersonic jet as the guest of the mayor. Mrs. Hughes enjoyed life fully until she died in her 114th year.

The Gompertz Law

The above optimistic view of the vitality of the oldest old is fully supported by recent statistical data on human mortality. Throughout most of life, human mortality steadily increases with age (called the Gompertz law of mortality). However, upon reaching the age of the oldest old, the rate of increase in mortality slackens and then stops altogether. In fact, the data for centenarians show that their mortality rate actually *decreases*.[29]

"After the age of 105 years, the one person in a million still alive has a greater chance of reaching age 106 than a person aged 104 has of reaching age 105."

---

[28]K. G. Manton, 1992, in *The Oldest Old*, (eds.), R. M. Suzman, D. P. Willis and K. G. Manton (Oxford University Press: Oxford), p. 157.

[29]R. E. Ricklefs and C. Finch, 1995, *Ageing* (Scientific American Library: New York), p. 6.

Kenneth Manton of the Center for Demographic Studies of Duke University, an authority in demographic patterns of ageing, has reached similar conclusions[30]:

> *"Vital statistics data show very little increase in mortality above age 100 ... There is a decline in mortality above age 104, based upon the British population registry data, which have extremely accurate reporting. Similar patterns have been found in the United States, Sweden and France."*

Conclusion

The conclusion that one may draw from all these recent scientific findings is that if one merits the traditional blessing of extremely long life, it is usually accompanied with good health. Divine favor is not marred by physical or mental infirmities.

---

[30] K. G. Manton, 1992, in *The Oldest Old*, (eds.), R. M. Suzman, D. P. Willis and K. G. Manton (Oxford University Press: Oxford), pp. 160–161.

Chapter 12

Problematic Numbers in Scripture II: The Extensive Population Growth of the Israelites in Egypt

Numbers

The Book of Numbers in Scripture is so named because this Book contains many numbers. Among them are numbers that are very difficult to understand. In this chapter, we shall deal with one such number and show how it can be explained.

Among the numbers that are difficult to understand is the enormous number of Israelites that left Egypt in the Exodus. Scripture states that 603,550 adult males left Egypt and this number does not include the tribe of Levi (Numbers 1:46). It is worth mentioning that when Scripture reports the number of members in a tribe, it includes only the males. One reason may be that the tribe of a child was determined by the tribe of his father and not his mother.

Although only 70 Israelites entered Egypt (Genesis, Chapter 46), when they left, they numbered over 600,000 adult males (Numbers 1:46). This number requires an explanation because it corresponds to a population growth of nearly 10,000, a population growth that is unprecedented in history.

This question is well known and various explanations have been proposed. We will show that the data recorded in Scripture, together with some reasonable assumptions, enable one to explain the vast population growth of the Israelites in Egypt.

Population Growth in History

The Israelites were in Egypt for 210 years. (The oft quoted number of 430 years refers to the period beginning with the birth of Isaac. See Rashi's commentary on Exodus 12:46.) The number of males leaving Egypt mentioned in Scripture includes only those males over the age of 20. Thus, we are speaking of a population increase during 190 years.

What is a typical population increase in history during a period of 190 years? For example, between the years 1740 and 1930, the world's population increased by only a factor of 3 (from 700 million to 2 billion). Thus, one must explain the enormous difference between such a modest increase in the world's population (only a factor of 3) and the enormous population increase recorded in Scripture (a factor of almost 10,000) over the same period of 190 years.

It is not possible to attribute the slow increase of the world's population to deaths caused by wars. The deadliest war in history was the Second World War, during which 60 million people died, including the six million Jews who were murdered. However, this enormous death toll had little effect (less than 3%) on the world's population.

The reason for the very slow increase in the world's population in previous times was the lack of knowledge of medicine. Before the modern age, the majority of children died young because of disease and plagues, and this is in addition to natural miscarriages and death during childbirth. Therefore, a large fraction of pregnancies did not produce a healthy child who lived to adulthood and ultimately had children of his/her own. Hence, the population of the world increased very slowly, despite the fact that the average woman became pregnant many times during her period of fertility. The situation is summarized in the famous expression: "Many pregnancies but small families."

In his prize-winning book, *Guns, Germs and Steel,* Jared Diamond, of the University of California at Los Angeles, describes the devastating effect that disease and plagues have had on various populations in the past. For example, at the time of the Exodus from Egypt, it took *a thousand years* for the world's population to double.

How Many Generations of the Israelites Were in Egypt?

As previously stated, the Israelites were in Egypt for 210 years. The fertility of a woman begins approximately at age 15 and continues to age 40–45. Any age between 40 and 45 years is equally reasonable to mark the end of a woman's period of fertility. We shall take the age of 45 years because this value leads to a result that is close to the population mentioned in Scripture. The length of a generation is taken to be a woman's age at the middle of her period of fertility, that is, 30 years, which lies midway between 15 years and 45 years. Therefore, *the 210 years that the Israelites were in Egypt correspond to seven generations.*

Size of a Typical Israelite Family

The main challenge is to determine the size of typical Israelite families in Egypt. We shall base our estimate on information given explicitly in Scripture. Jacob had twelve sons, but his family was not typical because he had four wives. Eleven sons of Jacob (not counting Levi) fathered 50 sons between them (Genesis, chapter 46). Some had many sons (Benjamin had 10 sons) and some had few sons (Dan had only one son). The average was 4.6 sons per family (50 divided by 11). We shall take this number of sons per family as the basis for our calculation. In addition to 4.6 sons per family, there were probably an equal number of daughters. Thus, we take the average family to consist of 9.2 children. There is nothing unusual about a family with nine children. Today, among very Orthodox Jews, having nine or more children is quite common.

In summary, the sons of Jacob (excluding Levi) sired 50 sons, and this number increased by a factor of 4.6 in every generation.

In the following table, we list the number of sons at the end of each generation (the numbers have been rounded off).

generation	beginning	1	2	3	4	5	6
males	50	240	1100	5060	23,300	107,000	492,000

The table does not list the sons born in the seventh generation, because at the time of the Exodus from Egypt, they were too young to be included in the census, which numbered only males "over the age of twenty." At the time of the Exodus, the first four generations had already died (120 years had passed). Therefore, we take into account only the fifth generation (107,000 males) and the sixth generation (492,000 males). The adult male population that left Egypt is thus calculated to be 599,000. This number is remarkably close to the figure of 603,550 adult males that appears in Scripture.

It is important to emphasize that we are not claiming that the details of the table describe exactly how many Israelites were living in each generation. Our purpose is to show that by combining the data given in Scripture with a few reasonable assumptions, one can explain the large number of Israelites who left Egypt at the time of the Exodus.

The Divine Blessing

It is written twice in Scripture that the Israelites in Egypt were blessed with a large increase in population, both in Exodus 1:7 (*"The Israelites were fruitful and multiplied and they became very numerous, and the land was filled with them"*) and also in Exodus 1:12 (*"The more that the Egyptians oppressed the Israelites, the more they increased in numbers"*). Since a family of nine children is not unusually large, how were these blessings expressed? *The meaning of the blessings is that most pregnancies resulted in children who lived to reach adulthood and to have children of their own.* This was highly unusual in the ancient world, and this was the expression of the divine blessings.

It is not necessary to assume that no Israelite children ever died young. Rather, from all the pregnancies of the average Israelite woman during her 30 years of fertility, nine pregnancies resulted in children who grew up to have children of their own. Thus, the Israelites increased in Egypt from a small tribe of seventy souls into a numerous and mighty nation.

The Difference Between the Tribe of Levi and the Other Tribes

The tribe of Levi requires a separate discussion for two reasons. First, although the increase in the number of the Levites seems impressive (from three males to 22,300 males), the tribe of Levi was significantly smaller than all the other tribes (Numbers, Chapter 4), whose numbers ranged from 35,400 (tribe of Benjamin) to 74,600 (tribe of Judah). And this resulted in spite of the fact that the recorded population of the other tribes included only males over the age of twenty (*"males from the age of twenty"*), whereas the recorded population of the tribe of Levi included babies (*"every male from the age of one month"*).

There is yet another reason for discussing the tribe of Levi separately. It is possible to carry out the same calculation for the tribe of Levi that was carried out for the entire population of the Israelites. However, it is not possible to do so for any other tribe because the *calculation requires the number of sons and also the number of grandsons* of that tribe. This information is given in Scripture only for the tribe of Levi. This enables one to calculate the increase of the tribe of Levi at the time of the Exodus from Egypt.

Moses and Aaron had special important tasks to fulfill during the Exodus from Egypt. Perhaps for this reason, Scripture found it appropriate to give detailed information about the family of Moses and Aaron, who were members of the tribe of Levi. Regarding the other tribes, there was no reason to present such detailed information. Therefore, it is not possible to calculate the expected increase in population of any other tribe.

Population Increase of the Tribe of Levi

The same calculation that we carried out for the entire population of the Israelites also explains the smaller number of Levites. It is written in Scripture that God commanded Moses to take a census of the tribe of Levi (Numbers 3:15):

> *"Count the Levites ... all males above the age of one month."*

The population of the tribe of Levi increased from the *three sons of Levi* (Gershon, Kohath, and Merari) at the time of entering Egypt to *22,300 Levites* at the time of the Exodus from Egypt (Numbers 3:39).

The relatively small number of Levites results from the smaller number of children in Levite families. According to Scripture, the typical size of the Levite family was 2.9 sons, whereas it was 4.6 sons for the other tribes (see previous discussion).

Size of a Typical Levite Family

Levi had three sons and eight grandsons. The sons of Gershon were Libni and Shimei, the sons of Kohath were Amram, Izhar, Hebron, and Uzziel, and the sons of Merari were Mahli and Mushi (Exodus 6:17–19). This corresponds to slightly fewer than three sons per family. This datum will serve as the basis of our calculation, that is, 2.9 sons per family, and, of course, an equal number of daughters.

The Number of Generations of Levites in Egypt

As previously stated, the Israelites were in Egypt for 210 years. The fertility of a woman begins approximately at age 15 and continues to age 40–45. Any age between 40 and 45 years is equally reasonable to mark the end of a woman's period of fertility. We shall take the age of 40 years because this value leads to a result that is close to the population of Levites mentioned in Scripture. The length of a generation is taken to be a woman's age at the middle of her period of fertility, that is, 27 years, which lies midway between 15 years and 40 years. Therefore, *the 210 years that the Israelites were in Egypt correspond to eight generations of Levites.*

In summary, three sons of Levi entered Egypt and this number of males increased by a factor of 2.9 in every generation. In the following table, we list the number of male Levites at the end of each generation (the numbers have been rounded off).

generation	beginning	1	2	3	4	5	6	7	8
Levites	3	8	23	70	200	580	1700	4900	14,200

At the time of the Exodus, the males of the first four generations of Levites had already died (over 100 years had passed). Therefore, we include the fifth generation (580 males), the sixth generation (1700 males), the seventh generation (4900 males), and the eighth generation (14,200 males). The total number of male Levites at the time of the Exodus from Egypt is thus calculated to be 21,380. This number is quite close to the number of males, 22,300, mentioned in Scripture.

Summary

Our goal here is to show that it is possible to explain the numbers mentioned in Scripture, even those that seem to be greatly exaggerated. The example before us was the extensive population increase of the Israelites in general and the tribe of Levi in particular. Our explanation is based on the data recorded in Scripture, combined with a few reasonable assumptions.

Chapter 13

Problematic Numbers in Scripture III: The Census of the Tribes

We here discuss the numbers connected with the two censuses of the Israelites mentioned in Scripture. The first census was taken shortly after the Israelites left Egypt (Numbers, Chapter 1), whereas the second census was taken forty years later (Numbers, Chapter 26).

A Whole Number of Hundreds

The task of the census was to determine the number of adult males in each tribe (with the exception of the tribe of Levi). Amazingly, *the population of adult males in 11 of the 12 tribes was reported to be a whole number of hundreds.*

There is a probability of only 1% that the population in two of the tribes would each be exactly a whole number of hundreds — a small probability but certainly possible. However, *the probability that the population of **each of 11 tribes** would be exactly a whole number of hundreds is less than one in a billion, billion!* In other words, the chance of this happening is completely negligible. The question becomes even more severe if one notes that this unbelievably rare event occurred twice — both in the first census (Chapter 1) and again in the second census (Chapter 26). *The probability of this event occurring **twice** is only one in a billion, billion, billion, billion!* How could this happen?

A Proposed Solution and Its Difficulty

This question is not new. One proposal is to give a different interpretation to the words "thousands" and "hundreds" that appear in Scripture regarding the census. The purpose of the census appears to be the counting of the number of soldiers in each tribe ("*every male twenty years and older, who was fit for military service*"). Therefore, perhaps the words "thousands" and "hundreds" in Scripture refer to military units. According to this proposal, "thousands" means "battalions" and "hundreds" means "companies," with each battalion consisting of ten companies.

For example, the results of the census of the tribe of Reuben were the following (Numbers 1:21):

> "*The tribe of Reuben numbered forty-six thousand and five hundred.*"

According to this proposal, the above verse means that the tribe of Reuben contributed 46 battalions and five companies to the Israelite army. Since Scripture is discussing the army, it is quite natural to enumerate each tribe in terms of the number of battalions and companies in that tribe, rather than in terms of the actual number of men. This proposal is supported by the fact that the census only counted males "*above the age of twenty*" (1:18), which would be the appropriate age for military service.

This proposal also explains the population of the tribe of Gad, which was *not* a whole number of hundreds: "*forty-five thousand, six hundred and fifty*" (1:25). This number means that in the tribe of Gad, in addition to 45 battalions and six companies, there was also a special unit of 50 men, perhaps a commando unit or an intelligence unit.

However, there is a difficulty with this proposal. In the first census of the tribe of Levi (Chapter 3) — a tribe that did *not* participate in the fighting — the population of Levites was also a whole number of hundreds. Moreover, this was the case for each family of Levites — the Gershonites (7500), the Kohathites (8600), and the Merarites (6200). Also in the second census of the tribe of Levi

(Chapter 26), the total number of Levites was a whole number of hundreds (23,000). Therefore, it is difficult to accept the proposal that Scripture is speaking about military units and not about numbers of men.

An Alternative Proposal: Rounded-Off Numbers

A more plausible explanation is that the numbers given in Scripture were always rounded off. That is, the number of adult males in the tribe of Reuben was *approximately* 46,500. Scripture has no interest in listing the exact number of men in each tribe. In fact, rounding off numbers is quite common in listing populations. According to this proposal, there is no difference between the census of the tribe of Levi and the census of the other tribes. The number of Levites was also rounded off to the nearest hundred.

Whenever Scripture mentions a number of several hundred or several thousand, the number was always rounded off to the nearest hundred or to the nearest thousand. Consider the following examples.

- When the messengers whom Jacob sent to Esau returned, they reported, *"We came to your brother Esau, and he is coming to meet you with 400 men"* (Genesis 32:6). This verse is not stating that the number of men was exactly 400, and not 399.
- Regarding the rebellion of Korah and the Israelites who were later punished, it is written: *"There arose against Moses ... 250 leaders of the community"* (Numbers 16:2), and *"The number who died in the plague was 14,700"* (17:14). Scripture is not stating the exact number of community leaders who rebelled or the exact number of Israelites who died. Both numbers have been rounded off.
- The punishment of the Israelites who worshiped idols in Shittim was the following: *"The number who died in the plague was 24,000"* (25:9). This verse is not stating that the number who died in the plague was exactly 24,000, rather than 24,001.

- In the Book of Esther, it is written: "*In the capital Shushan, the Jews killed 500 men*" (Esther 9:6), and: "*In the provinces … they killed 75,000 of their enemies*" (9:16). Once again, the numbers were rounded off. The first number is given to the nearest hundred and the second, larger number is given to the nearest thousand.
- There is a famous question regarding a vessel in the palace of King Solomon. The First Book of Kings states (7:23) that its diameter was 10 cubits and its circumference was 30 cubits. Already in the time of Solomon, it was known that the circumference of a circle is slightly more than 3.14 times its diameter, a ratio that mathematicians denote by the Greek letter π. Therefore, why is the circumference of the vessel given as 30 cubits, whereas it was actually larger than 31 cubits?

The commentator Levi ben Gershon (fourteenth century) explains that the value for the circumference given in the First Book of Kings is not exact but it was rounded off: "*The stated circumference of 30 cubits is **an approximate value**.*"

Summary

From the above examples, one sees that throughout Scripture, large numbers were rounded off if there is no specific importance to the exact number. This practice follows the well-known saying: "*Scripture employs the language used by people.*"

Chapter 14

String Theory, Ten-Dimensional Universe, and Kabbalah

In recent years, many religious scientists, I among them, have written about the emerging harmony between the discoveries of modern science and Scripture, in particular, the Genesis account of the creation of the universe.[1] The Big Bang theory of cosmology provides a scientific explanation[2] for every word and every phrase that appear in the first five verses of Genesis — the First Day of Creation. In view of these striking correlations between science and Scripture, it is tempting to explore a later traditional source that discusses the creation of the universe, namely, Kabbalah — the book of Jewish mysticism.

There are various traditions in Kabbalah. Our presentation will follow the ideas of Isaac Luria (sixteenth century), whose approach to Kabbalah follows the writings of earlier scholars of mysticism. The ideas of Luria were written down by his disciple, Hayim Vital.[3]

---

[1] N. Aviezer, 1990, *In the Beginning* (Ktav Publishing House: New York); J. Landa, 1990, *Torah and Science*; G. Schroeder, 1991, *Genesis and the Big Bang*; Y. Levi, 2004, *Science in Torah*.

[2] This statement relates to words that have *physical* content. However, there are words in the first chapter of Genesis that have only *spiritual* content, such as "*The spirit of God hovered over the surface of the water*" (1:2). Clearly, science cannot tell us anything about the meaning of this sentence.

[3] Hayim Vital, *Etz Hayim*.

The creation narrative that appears in Kabbalah is very different from the creation narrative in the first chapter of Genesis. This does not imply any contradiction between these two accounts of the same event. Rather, the two versions of creation emphasize different features. Genesis deals with the actual sequence of events (First Day, Second Day, etc.), whereas Kabbalah stresses the role of God in the process of creation.

Is it possible to correlate the creation narrative given in Kabbalah with the findings of modern science? One might object to this question on the grounds that Kabbalah deals with the spiritual realm, whereas science is restricted to the physical realm. Nevertheless, it is a principle of Kabbalah that the spiritual realm of the "world above" descends, suitably garbed, to form a physical counterpart in the "world below." Therefore, it is in place to ask: Are there features of the physical world ("the world below") that are related to Kabbalah ("the world above")? We shall see that the answer to this question is: "Yes."

In the past few decades, the physical universe has been revealed to be a far richer, more complex and wonderful place than anyone could have imagined. It is precisely this recently discovered subtlety and intricacy of the universe that provides the framework for the various correlations with the spiritual world of Kabbalah.

Kabbalah

There are learned scholars who spend a lifetime studying the mysteries of Kabbalah. It is obvious, therefore, that this chapter will not contain a comprehensive account of the subject. For our purposes, it is sufficient to deal with a few basic concepts.

One of the basic concepts in Kabbalah is the *sefirot*, which is analyzed in detail in books about Kabbalah. This term has been understood variously as referring to "spheres" or to *sapirim* that "radiate and sparkle" or to *mesaprim* that "relate" the glory of God. The essence of God cannot be known. We know of God only through

His manifestations. Central to His manifestations are the ten *sefirot*, which represent divine emanations or dimensions.

Kabbalah Account of Creation

Kabbalah characterizes God as the *Ein-Sof* ("infinite"), a limitless and unknowable infinite realm. The ten *sefirot* are configurations of divine powers within the Godhead, containing the principles whereby God manifests Himself to us. They constitute the means through which God interacts with the universe.

In the beginning, the universe did not exist. The existence of an entity in addition to the *Ein-Sof* would have constituted a limitation on His infinity. To enable the universe to exist required an act of *tzimtzum* on the part of God. The literal meaning of *tzimtzum* is "contraction," which Luria explains as "withdrawal." This divine withdrawal made possible the creative processes leading to an entity — the universe — that could exist in parallel with the *Ein-Sof.*

God's withdrawal provided a space into which flowed a ray (*kav*) of divine light. From this divine light, the entire universe developed. Further details of this light are described in the Kabbalah literature. What is relevant to our discussion is the effect of this light on the *sefirot* or, more accurately, on the "vessels" (*kelim*) which enveloped each of the ten *sefirot.*

The vessels of the first three *sefirot* managed to contain the ray of light that flowed into them. However, as the light struck the following seven *sefirot*, it became too powerful to be held by their vessels, which cracked and shattered, one after another.

This Kabbalah concept is known as "the breaking of the vessels" (*shevirat ha'kelim*).

Scientific Account of Creation

The branch of science that deals with the origin of the universe is known as cosmology.

In every age and in every culture, people would look up at the sky and wonder: What was the origin of the heavenly bodies — the sun, the moon, and the stars? Throughout the ages, scientists considered *creation* to be impossible because it was assumed that *something* cannot come from *nothing*. Therefore, scientists viewed the universe as eternal, thus avoiding questions regarding its origin. The statement in Genesis that the universe was *created* became an arena of conflict between science and Scripture. That is how matters stood for many years.

This situation has now changed. The twentieth century witnessed an explosion of scientific knowledge, which was especially dramatic in cosmology. Astronomers had been studying the heavenly bodies for thousands of years, but their studies dealt almost exclusively with charting the paths of the stars, planets, and comets, and determining their composition, spectrum, and other properties. The *origin* of the heavenly bodies remained a complete mystery. Important advances in cosmology during the past few decades have now permitted scientists to construct a coherent history of the origin of the universe. Today, an overwhelming body of scientific evidence supports the "Big Bang" theory of cosmology.[4]

The scientific status of the Big Bang theory was summarized in the prestigious journal *Scientific American* as follows[5]: *"The Big Bang theory works better than ever."* Similarly, Brian Greene, of Columbia University, writes[6]:

> *"The modern theory of cosmic origins asserts that the universe erupted from an enormously energetic event. The Big Bang theory of creation is referred to as the standard model of cosmology."*

---

[4] There are four major pieces of evidence: (1) the discovery of the remnant of the initial ball of light that fills the universe, (2) the hydrogen-to-helium ratio in the universe, (3) the expansion of the galaxies, and (4) the perfect black-body spectrum of the microwave background radiation measured by the COBE space satellite in 1989, and the further measurements of this radiation made by the WMAP space satellite launched in 2001.

[5] G. Musser, February 2004, *Scientific American*, p. 30.

[6] B. Greene, 1999, *The Elegant Universe* (Jonathan Cape: London), pp. 345–346.

The central theme of the Big Bang theory is that *the universe began through an act of creation.* It is instructive to quote some of the world's foremost authorities on the subject of creation.

Nobel laureate Paul Dirac, a major architect of twentieth-century physics, writes[7]:

> *"It seems certain that there was a definite time of creation."*

Leading cosmologist Stephen Hawking writes[8]:

> *"The creation lies outside the scope of the known laws of physics."*

When cosmologists use the term "creation," to what are they referring? Precisely what object was created? Scientists have discovered that the universe began with the sudden appearance of an enormous *ball of light,* called the "primeval light-ball" and later dubbed the "Big Bang" by the British astrophysicist Fred Hoyle. The remnant of the primeval light-ball was detected in 1965 by two American scientists, Arno Penzias and Robert Wilson, who were awarded the Nobel Prize in Physics for their discovery.

In the following years, additional scientific evidence supported the Big Bang theory. In particular, the data measured by the satellites COBE in 1989 and WMAP in 2001 were in precise agreement with the predictions of the Big Bang theory.

Comparing Kabbalah with Science

The principal features of the Kabbalah account of creation are the following:

- The universe began through an act of creation.
- Divine light played a central role in the creation.
- There exist three "intact" *sefirot* and seven "broken" *sefirot.*

---

[7] P. Dirac, 1992, *Commentarii,* vol. 2, no. 11, p. 15.
[8] S. Hawking, 1973, *The Large Scale Structure of Space-Time* (Cambridge University Press: Cambridge), p. 364.

We will now relate these three features of the Kabbalah account of creation to the scientific account of the creation of the universe.

The first feature of the Kabbalah account deals with an *event* — the creation.

The Big Bang theory of cosmology asserts that the universe was indeed created. Today, one cannot carry on a meaningful discussion of cosmology without the creation of the universe assuming a central role.

The second and third features of the Kabbalah account of creation deal with *entities* — the divine light and the ten *sefirot*. According to Kabbalah, every entity of the spiritual "world above" descends, suitably garbed, into the physical realm of the "world below." Therefore, one seeks in the "world below," the physical counterparts to the divine light and the ten *sefirot*.

The physical counterpart of the divine light of Kabbalah is clearly the primeval light-ball of the Big Bang. This theory of cosmology asserts that the entity that was created at the beginning of time was an enormous ball of light, popularly known as the "Big Bang" and hence the name of the theory. With appropriate instrumentation, one can still observe the remnant of this primeval light that dates back to the very origins of time.

The difficulty resides in the third feature of the Kabbalah account of creation.

What might be the physical counterpart of the ten *sefirot*? Since the *sefirot* are often viewed as the "dimensions of God," we propose that the physical counterparts of the *sefirot* are the *dimensions of the universe*. Accordingly, the three "intact" *sefirot* correspond to the three familiar dimensions of space: east–west, north–south, and up–down.

This brings us to the crux of the problem.

- The total number of *sefirot* is ten. Can there be a ten-dimensional universe?
- Seven of the *sefirot* were broken. Is there such a thing as a "broken" dimension?

Interestingly, both these questions are addressed by the latest scientific findings regarding the force of gravity.

Gravity

A ten-Dimensional Universe and Kabbalah

Over the years, the theory of gravity has undergone a number of important changes.

According to the modern theory of gravity, the universe does indeed consist of ten dimensions. This is the basis for the relationship between gravity and the ten divine dimensions of Kabbalah.

a. **Newton's theory of gravity**: The first theory of gravity was formulated in 1687 by Isaac Newton in his *Principia,* one of the most important books of science ever written. Newton introduced the idea that every two objects in the universe attract each other with a force, called gravity, whose magnitude depends on the distance between the objects. This theory enabled Newton to explain planetary motion, as well as many other phenomena, such as the tides.

b. **Einstein's theory of relativity**: In 1905, Albert Einstein proposed the special theory of relativity, establishing the relationship between matter (M) and energy (E) through his famous formula $E = Mc^2$, where the letter c denotes the speed of light. Einstein's theory of special relativity has been confirmed countless times and is now one of the fundamental principles of science. Every scientific theory *must* be compatible with the theory of special relativity. However, Newton's theory of gravity is not compatible with relativity. Therefore, it was realized that Newton's theory of gravity had to be modified.

c. **Einstein's theory of gravity**: Einstein struggled for a decade to formulate a new theory of gravity that *was* compatible with special relativity. In 1915, he announced his theory of gravity, which he called the "general theory of relativity."

The most surprising result of Einstein's theory is that *gravity is not a force*. Rather, gravity is a "distortion of space."[9] The gravitational attraction between two objects is *not* due to one object pulling the other object, as is the case for the electric force between two electric charges. Instead, the first object "distorts" the space around it, and the second object moves *in reaction to this distortion of space*. Since the distortion of space is not visible, it appears to us *as though* the two objects are attracted to each other by means of a force.

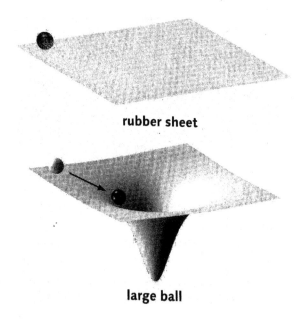

rubber sheet

large ball

The concept of a distortion of space can best be explained by means of the accompanying figure. The top part of the figure shows a stretched rubber sheet on which a small ball lies motionless. The bottom part shows the situation after a large ball has been placed on the rubber sheet. The effect of the large ball is to distort the rubber sheet, with the distortion being greatest in the vicinity of the large ball.

---

[9] Einstein's theory of gravity also implies a *distortion of time* (called "time dilation") since time is linked to space. However, we shall not be discussing *time* here since the subject of time in Kabbalah would require a separate article.

As a result of the distortion of the rubber sheet, the small ball begins to move toward the point of maximum distortion (which is where the large ball lies). In other words, the small ball moves *towards* the large ball in spite of the fact that *there is no force of attraction between the two balls.* The motion of the small ball is caused by the distortion of the rubber sheet, which in turn is caused by the large ball.

The "rubber sheet" represents space whose "distortion" is invisible to us. We see only the small ball moving towards the large ball. Therefore, Isaac Newton assumed that a force of attraction (gravity) exists between every pair of masses. However, Einstein demonstrated that the correct description of gravity is a distortion of space (as illustrated by the rubber sheet and its distortion) rather than a force.

The theories of gravity proposed by Einstein and by Newton are fundamentally different. In practice, however, the *predictions* of these two theories are very similar.

In fact, their predictions are so similar that for two centuries, no one doubted Newton's theory. However, whenever the two theories do differ in their predictions, it is always Einstein's theory that agrees with the detailed observations. Therefore, Einstein's theory of gravity is today accepted by all scientists.

In summary, gravity is not a force, but rather a distortion of space. Thus, gravity is totally unlike the other forces of nature. We shall soon see the importance of this.

d.　**Quantum theory**: The twentieth century witnessed the development of quantum theory. (The best layman exposition of quantum theory is Richard Feynman's book, *QED, The Strange Theory of Light and Matter.*) Quantum theory has been confirmed by thousands of experiments, and it has become a basic principle of physics. Therefore, every theory of physics *must* be compatible with quantum theory.

Feynman was awarded a Nobel Prize for developing a procedure, known as "renormalization," for making every theory of forces (the electromagnetic force and the two nuclear forces) compatible with quantum theory. However, Feynman's procedure cannot be applied to Einstein's theory of gravity because according to Einstein, gravity is not a force, but rather a distortion of space. It is this feature that makes Einstein's theory of gravity incompatible with quantum theory.

This is a very serious problem. Quantum theory is certainly correct and Einstein's theory of gravity is certainly correct. How can two correct theories be incompatible with each other? In order to make these two theories compatible, a new scientific framework for describing the universe had to be developed. A theory of gravity that is compatible with quantum theory is called "*quantum gravity*".

e.   **String theory:** The new scientific framework that resolves the incompatibility between quantum theory and Einstein's theory of gravity is known as string theory.[10] This theory is a revolutionary conceptual framework for describing the physical universe. According to the previous conception, the basic entities of the universe are particles — electrons, quarks, photons, etc. String theory asserts that the basic entities of the universe are not particles, but rather, they are tiny "strings." These strings vibrate (like a violin string) and the energy of vibration *appears* as a "particle" through the Einstein relation between energy and mass, $E = Mc^2$. (For an excellent layman account of string theory, see Brian Greene's book, *The Elegant Universe*.)

String theory can be formulated for various dimensions. In three dimensions, string theory fails to yield a theory of quantum gravity. Moreover, for any number of dimensions of space *fewer than ten*, string theory fails to yield a theory of quantum gravity. However, *for a ten-dimensional universe, string theory does yield a consistent theory of quantum gravity*.

---

[10] String theory has gone by a variety of names during its development, including superstring theory, brane theory, and *M*-theory, with the last designation being popular among the *cognoscenti*. We shall here use the more widely known name of "string theory," even though we shall be quoting the latest results of *M*-theory.

*In summary, scientists discovered that Einstein's theory of gravity is compatible with quantum theory **only** in the framework of string theory and **only** if the universe contains ten dimensions of space. These two theories must be compatible with each other. Hence, scientists conclude that the universe must have ten spatial dimensions.[11]*

f. **Connection with Kabbalah**: The combination of Einstein's theory of gravity with string theory implies that the universe consists of ten dimensions. Thus, the ten dimensions of the universe may be interpreted as the physical counterpart of the ten *sephirot* of Kabbalah.

Compacted Dimensions and Kabbalah

We now turn to the other feature of the *sephirot* of Kabbalah that requires an explanation, namely, the seven "broken" *sephirot*. What aspect of the dimensions of space can correspond to a "broken" dimension?

How can one reconcile the ten-dimensional universe of string theory with our everyday experience of a three-dimensional universe? What is the meaning of the "missing" seven dimensions?

The three familiar dimensions of space (up-down, east-west, north-south) are infinite in extent. The other seven dimensions are so very short that they are not accessible to our senses. For this reason, it was previously thought that we inhabit a universe of only three dimensions.

The seven very short dimensions are said to be "compacted." According to string theory, the extent of a compacted dimension is a billionth of a billionth of a billionth of the radius of an atom

---

[11] The cover of the November 2003 issue of *Scientific American* declares: "String Theory and Space-Time with Eleven Dimensions." The eleven dimensions of space-time consist of one dimension of time and the ten dimensions of space that are posited by string theory. Einstein's theory of special relativity asserts that space and time are interwoven, and one therefore speaks of *space-time*. However, we here restrict the discussion to *spatial* dimensions. A discussion of the concept of *time* in Kabbalah would require an essay all by itself.

(called the "Planck length," in honor of Nobel laureate Max Planck). Such a short dimension can never be detected with any measuring device. This is the meaning of "compacted dimensions" — dimensions that exist but are far too short to be measured.

STRING THEORY

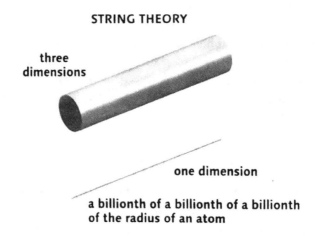

three
dimensions

one dimension

a billionth of a billionth of a billionth
of the radius of an atom

The concept of compacted dimensions is illustrated in the accompanying figure. The top part of the figure shows a rod, whose three dimensions are its length and its cross-sectional area. In the bottom part of the figure, the cross-sectional area of the rod is so diminished that the rod has become reduced to a thin wire. As the cross-sectional area is further diminished, the two dimensions of the thin wire become observable only with difficulty, and eventually, they cannot be observed at all.

Even though compacted dimensions cannot be measured *directly*, they have very significant *indirect* effects on the universe. The most important of these indirect effects is that the seven compacted dimensions lead to a theory of quantum gravity.

Kabbalah, Science, and Creation — Summary

String theory is the modern scientific framework for understanding the universe. The most amazing prediction of string theory is that the universe consists of ten dimensions. Of these ten dimensions, three are the familiar infinite dimensions of space, while the remaining seven dimensions have become "compacted" during the process of creation.

Kabbalah speaks of ten *sefirot* which describe the spiritual "world above." Of the ten *sefirot* of the "world above," three remained "intact" while the other seven became "broken" during the process of creation. We identify the ten dimensions of the physical universe, the "world below," as the counterpart of the ten *sefirot* of the spiritual "world above." The three intact *sefirot* correspond to the familiar three infinite dimensions of space. The seven *compacted* dimensions of the "world below" are the counterparts of the seven *broken sefirot* of the "world above."

Thus, there is a correspondence between the Kabbalah account of the creation of the universe and the scientific account of creation, which is based on the Big Bang theory of cosmology, Einstein's theory of gravity, and string theory.

Chapter 15

Scripture and the Spread of Languages

The Tower of Babel and the Noahide Languages

One of the most interesting events in Genesis is the incident of the Tower of Babel.

Scripture relates (Genesis 11:1–9) that at one time everyone spoke the same language and they assembled to build a mighty structure — the Tower of Babel. God was very displeased with this plan, and He dispersed the people and confounded their language. This population dispersal led to the development of different nations and gave rise to the various languages of the ancient world.

The above account of the Tower of Babel offers no details about which peoples and languages developed as a result of the dispersal of the population. This information is given in Chapter 10 of Genesis, which is devoted to the genealogy of Noah's three sons: Shem, Ham, and Japheth. Each of Noah's sons is described as the progenitor of more than ten nations,[1] and a list is given for each son. Of specific interest to the present discussion is the fact that the dispersal of Noah's descendants did not simply result in separate peoples, but also led to the development of separate languages:

---

[1] The term "nation" is not intended to imply a modern nation-state, but a chiefdom or tribal society.

"From these [descendants of Japheth]...each according to his language" (10:5).

"These are the descendants of Ham...according to their languages" (10:20).

"These are the descendants of Shem...according to their languages" (10:31).

Scholars do not agree on the exact geographical location of the various Noahide languages, but such differences in detail are not relevant for our discussion. The locations of the Noahide languages given in the *Hammond Atlas of the Bible*[2] are similar to those given in *Carta's Atlas of the Bible*.[3]

We shall here compare the development of the Noahide languages, as related in Scripture, with the latest findings in linguistic research. In contrast to a widespread misconception, we shall see that the Genesis text is, in fact, in agreement with recent discoveries in comparative linguistics.

The Languages of Shem and Ham

The Semitic and Hamitic languages, attributed to the descendants of Noah's sons Shem and Ham, are all related. Linguists classify them as belonging to the Afro-Asiatic family of languages (in the linguistic context, "Afro" means North African and "Asiatic" means Middle Eastern).[4] The Afro-Asiatic languages of the ancient world included Hebrew, Assyrian,[5] Egyptian, Babylonian, Aramaic, Amorite, Moabite, and Cushite. All these names are familiar from the Book of Genesis.

---

[2] C. S. Hammond, 1959, *Atlas of the Bible* (Hammond: New York), map on p. B-4.

[3] Y. Aharoni, 1974, *Carta's Atlas of the Bible* (Carta: Jerusalem), map 15 on p. 21.

[4] D. Crystal, 1997, *The Cambridge Encyclopedia of Language*, 2nd edition (Cambridge University Press: Cambridge), p. 318.

[5] The Assyrians spoke the ancient Semitic language known as Akkadian. There were two dialects of Akkadian, the Assyrian dialect spoken in northern Mesopotamia and the Babylonian dialect spoken in southern Mesopotamia. For this reason, the Akkadian language is also referred to as Assyro-Babylonian.

It is the Japhethide languages that raise difficult questions about the Genesis account of the spread of languages. These will be the subject of this chapter, and it will be seen that Scripture and modern linguistics are, in fact, in close agreement.

The Languages of Japheth

The most interesting feature of the Japhethide languages is their vast geographical extent.

As the accompanying map shows, the Japhethide languages listed in Genesis were spoken throughout Europe and deep into Asia. The European branch of the Japhethide languages extends from Greece through Germany and as far west as Spain, whereas the Asian branch extends from Persia to the ancient Kingdom of the Medes (present-day Northern Iran and Afghanistan) and as far east as ancient India (present-day Pakistan).

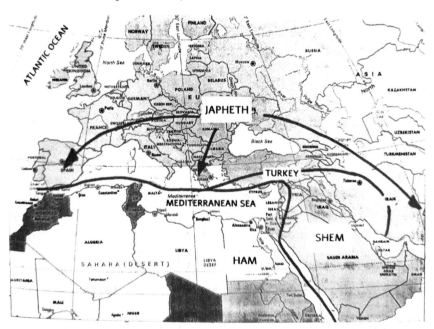

The Genesis account of Japhethide languages implies the following:

- There should be a linguistic relationship between the various Japhethide languages, including ancient Greek and German (European) and ancient Persian and Indian (Asian).
- The Japhethide languages should show signs of having originated near Turkey, since they all developed during the dispersal of Noah's descendants after the Flood. Noah's ark landed on Mount Ararat, which is in eastern Turkey.
- The most ancient of these languages should have originated around the date of the Flood, about 4000 years ago.
- The languages of Japheth did not spread by conquest. Genesis implies that the descendants of Japheth developed new languages as they peacefully migrated into previously unoccupied lands in Europe and Asia.

We shall see that all these statements agree with current findings in linguistics.

European Languages

The classification of the world's languages is a major area of linguistics research. Scholars compare various languages, seek correspondences between them, and group related languages into families.

There are about 6000 languages spoken in the world today. Most widespread are the languages of Europe. Although European languages comprise only 3% of the world's languages, they are the native tongue of nearly half the world's population.[6]

It has long been obvious that almost all the European languages are related. French, Italian, and Spanish all stem from Latin. The

---

[6] D. Crystal, 1997, *The Cambridge Encyclopedia of Language*, 2nd edition (Cambridge University Press: Cambridge), pp. 286–287.

Scandinavian languages are very similar. German, Dutch, and English share many words, as do all Slavic languages. Careful studies of vocabulary, syntax, and phonology have established that, with four exceptions,[7] *all* the languages currently spoken in Europe belong to one single family.

However, as one leaves Europe, the situation changes radically. The languages of the neighboring countries of Turkey, Georgia, and the Middle East are unrelated to those of Europe.

The Problem

In the eighteenth century, linguistic opinion was at variance with the Genesis account of the Japhethide languages. At that time, linguists believed that there is *no connection* between the ancient languages of Asia and those of Europe. By contrast, the Japhethide languages include *both* Asian languages (e.g., ancient Persian and Median) *and* European languages (e.g., the ancient languages of Germany and Greece). Moreover, between the Asian Japhethide languages and Europe lie the countries of Turkey, Georgia, Azerbaijan, Iraq, and Syria. In none of these intervening countries is a European language spoken.

These facts clearly present a problem, because they seem to be inconsistent with the statement in Genesis that all the Japhethide languages are related.

The Indo-European Family of Languages

Sanskrit: A Linguistic Surprise

One of the most important events in the history of linguistics was the shattering of the belief that there is no relationship between

---

[7]Finnish, Hungarian, and Estonian are Uralic languages from Asia, whereas the Basque language, spoken in parts of northeast Spain, is an "isolate," that is, unrelated to any other known language.

European and non-European languages. The linguistic bombshell fell in 1786, when Sir William Jones, an English oriental scholar serving as a judge in India, made an extraordinary discovery.[8] While in India, Jones took up the study of Sanskrit, the extinct language of the early literary and religious texts of India. Even after it was no longer spoken, Sanskrit continued to be the language of scholarship and literature, similar to the role of Latin in the West in the Middle Ages.

In his "Third Anniversary Discourse" to the Asiatic Society of Bengal, Jones made the following observations regarding Sanskrit[9]:

> *"The Sanskrit language has a wonderful structure, more perfect than Greek and more copious than Latin. It bears to both of them a strong affinity, both in the roots of verbs and in the forms of grammar ... so strong that no philologist could examine these three languages without believing them to have sprung from acommon source. Also Gothic [ancient German] and Celtic had the same origin as Sanskrit, and old Persian should be added to the same family."*

Jones recognized that a deep connection exists between the ancient languages of Europe and those of Asia. This brilliant observation was subsequently developed by many other linguists. Finally, in 1813, the English scholar Thomas Young introduced the term "Indo-European" for this widespread family of languages. Moreover, the Asian branch of the Indo-European languages is not restricted to Sanskrit, ancient Persian, and a few other languages. Asia is the homeland of over 40% of the Indo-European languages.

---

[8] D. Crystal, 1997, *The Cambridge Encyclopedia of Language*, 2nd edition (Cambridge University Press: Cambridge), p. 288.

[9] W. Jones, 1807, *The Collected Works of Sir William Jones III* (John Stockdale: London), p. 23.

The Language of Ancient Turkey: Another Surprise

An important discovery in the study of Indo-European languages was the deciphering of the language of ancient Turkey. The course of events has been described as follows[10]:

In the late nineteenth century, excavations in Turkey uncovered thousands of tablets in an unknown language. These tablets remained a mystery until 1917, when scholars were astonished to find that this language belonged to a previously unknown ancient branch of the Indo-European family. This language, called Anatolian, is the earliest Indo-European language discovered to date.

The Discrepancy Resolved

Today, linguists recognize that the Indo-European family of languages links ancient Asian languages with ancient European languages, in precise accord with the Genesis list of Japhethide languages. Moreover, in Turkey, the ancient Anatolian language (also known as Old Hittite) was also found to be an Indo-European language. (In modern Turkey, an unrelated Altaic language is spoken.) Therefore, we see that *all the Japhethide languages belong to the same Indo-European family of languages*, in agreement with the description given in Genesis.

The Ancestral Homeland of the Indo-European Languages

Once it became clear that Indo-European languages are found in both Asia and Europe, scholars naturally began to wonder about the origin of this widespread linguistic family.

The proposals included the following. The Indo-European languages originated in Asia and then spread westward into Europe? Or they originated in Europe and then spread eastward into Asia?

---

[10] J. Diamond, 1991, *The Rise and Fall of the Third Chimpanzee* (Vintage: London), p. 236.

Or perhaps they had originated near the European–Asian border, say, in Turkey, and then spread *both* eastward into Asia *and* westward into Europe? The last scenario corresponds to the account given in Genesis.

The effort to locate the original homeland of the Indo-European languages has been called the Indo-European Problem, and it has occupied generations of linguists.

The deep similarities between the various Indo-European languages clearly indicate that all of them derived from a single ancestral language, older than Sanskrit, Greek, or Latin. This ancestral language, called "Proto-Indo-European" or PIE, was reconstructed by studying cognate words (words of common origin) in various Indo-European languages. For example, a comparison of the English word *birch*, the German *birke*, the Lithuanian *berzas*, the Old Slavonic *breza*, and the Sanskrit *bhurja* indicates that there existed a parent word for birch tree in PIE. From the vocabulary of PIE that was thus constructed, linguists developed a picture of the world inhabited by its original speakers and their environment before their dispersal from their original homeland.

From such studies, it is possible to estimate when PIE was last spoken. Jared Diamond, of the University of California, emphasizes that the words that appear in PIE, and, even more importantly, the words that are absent in PIE, serve as an indication of which items were used by these ancient people. The absence of a word indicates that the object in question was unknown to the speakers of PIE. For example, there is no word in PIE for "iron," which suggests that iron was unknown until after the breakup of PIE into various daughter languages. Combining these results with archaeological evidence about the time when various items first came into use, Diamond has estimated that PIE began to develop into daughter languages about 4500 years ago.[11]

We note that this date estimated by Diamond corresponds closely with the date of the biblical Flood.

---

[11]J. Diamond, 1991, *The Rise and Fall of the Third Chimpanzee* (Vintage: London), pp. 237–238.

Such considerations are also helpful in determining the location of the ancestral homeland of PIE speakers. Over time, many possible homelands have been proposed, including Central Asia, Northern Europe,[12] Central Europe, and north of the Black Sea.[13]

A New Proposal for the Ancestral Homeland of PIE

Colin Renfrew, Professor of Archaeology at the University of Cambridge, is a leading authority on Indo-European languages. In his book, *Archaeology and Language: The Puzzle of Indo-European Origins*, Renfrew stresses the importance of archaeological evidence in determining the ancestral homeland of the Proto-Indo-Europeans. Critically reviewing the various proposals and explaining their short-comings, Renfrew asserts that "these proposals do not provide the solution to the Indo-European problem."[14]

Renfrew then marshals the evidence in favor of his new proposal that the first speakers of Proto-Indo-European lived in Turkey. He concludes as follows[15]:

---

[12] The search for the proto-Indo-European homeland has not always remained an academic discussion. For most of the twentieth century, German scholars favored an ancestral homeland in Northern Europe, and they referred to the original PIE speakers as Aryans, a term introduced by Gordon Childe in his 1926 book *The Aryans: A Study of Indo-European Origins*. The Nazis perverted this idea into the concept of a master race of pure-blooded Aryans who had lived in ancient Germany (Northern Europe) and who were responsible for everything worthwhile in Western civilization. The Nazis regarded themselves as the twentieth-century descendants of the Aryans, assuming for themselves the role of guardians of the "purity" of Western civilization, with the task of destroying all non-Aryan influences. The result was the Holocaust.

[13] D. Crystal, 1997, *The Cambridge Encyclopedia of Language*, 2nd edition (Cambridge University Press: Cambridge), p. 298.

[14] C. Renfrew, 1987, *Archaeology and Language: The Puzzle of Indo-European Origins* (Jonathan Cape: London), p. 98.

[15] C. Renfrew, 1987, *Archaeology and Language: The Puzzle of Indo-European Origins* (Jonathan Cape: London), p. 205.

"Central and eastern Anatolia [present-day Turkey] was the key area where the early form of Indo-European was spoken. From there, the spread of the language and its successors into Europe was associated with the spread of farming."

In an article, entitled "The Origins of Indo-European Languages," Renfrew writes[16]:

"Almost all European languages are members of a single family of languages which spread by peaceful diffusion ... The traditional view holds that the ancestral PIE language was spread by nomadic horsemen who lived north of the Black Sea. These mounted warriors conquered indigenous peoples and imposed their PIE language, which eventually evolved into the European languages we know today. I here offer a different view, based on new insights. According to this view, the spread of the Indo-European languages did not require conquest. It was a peaceful diffusion from its origins in Anatolia [Turkey] and the Near East."

Current Status of Renfrew's Proposal

As of this writing, Renfrew's proposal is still the subject of debate. Some scholars continue to support the older, traditional view that the northern coast of the Black Sea was the ancestral homeland of the Indo-Europeans.[17] As Renfrew himself writes[18]:

"It would be wrong to assume that the last word on this topic has been spoken.

My proposal of an Anatolian origin of the Indo-European languages finds validation in recent research, but the final picture will no doubt be more complex.

When a more complete understanding is achieved, the spread of farming from Turkey into Europe will prove to be a significant part of the story."

---

[16] C. Renfrew, October 1989, *Scientific American*, p. 82.

[17] J. Diamond, 1991, *The Rise and Fall of the Third Chimpanzee* (Vintage: London), pp. 241–247.

[18] C. Renfrew, October 1989, *Scientific American*, p. 90.

The Words of Scripture

Having presented the linguistic evidence regarding the Indo-European family of languages, we return to the questions posed earlier regarding the Japhethide languages.

- The Genesis list of Japhethide languages is consistent with the Indo-European family of languages. These include many European languages, as well as the Asian languages of Persian, the language of the Kingdom of the Medes, and the ancient language of Turkey. All these appear on the Genesis list of Japhethide languages.
- The location of the original homeland of the Indo-European languages is a subject of debate. Colin Renfrew identifies Anatolia (present-day Turkey) as the place from which these languages spread throughout Europe and Asia. This location corresponds to the landing site of Noah's ark on Mount Ararat.

  A rival theory to that of Renfrew suggests the northern shore of the Black Sea as the place of origin of these languages. This locale is not far from Turkey, which lies on the southern shore of the Black Sea. Thus, the rival theory is also in reasonable agreement with Genesis.
- The date proposed for the beginning of the separation of Proto-Indo-European into daughter languages (about 4500 years ago) is close to that of Noah's Flood.

In conclusion, we find that current knowledge in linguistics is in agreement with the Genesis account of the languages of the descendants of Japheth.

> *"God said, 'Behold, they are one people with one language, and this is what they begin to do. What they propose should be withheld from them. Let us descend and confuse their language, so that they will not understand each other.' God then dispersed them from there over all the earth, and they stopped building the city. That is why the city was called Babel, because from there God confused their language, scattering them over the earth"* (Genesis 11:6–9).

Chapter 16

Quantum Theory, Photons, and Scripture

The Rainbow

In the Book of Genesis, Scripture relates the narrative of the Flood. Because of the immoral behavior of mankind, God caused an enormous flood that drowned all human beings, except for the righteous family of Noah which was saved in the Ark.

After the flood waters had receded and Noah and his family left the Ark, God promised that He would never again destroy the world by a flood and He gave a visual sign of His promise (Genesis 9:12–15):

> *"This is the sign of the covenant that I am establishing between me, you, and all living creatures forever. My rainbow placed in the cloud will be the sign of the covenant between me and the Earth. When there will be clouds on the Earth and the rainbow will appear in the clouds, I will remember the covenant between me, you, and all living creatures, and there will never again be flood waters to destroy all creatures."*

These verses seem to imply that the rainbow first appeared after the Flood as a visual sign of God's promise that He would never again cause a flood to destroy all mankind. However, the rainbow, the visual sign of God's promise, is known to be a natural phenomenon that accompanies the rain. Sunlight is a combination of all colors. When it rains, the raindrops act as prisms that separate sunlight into its various colors — the colors of the rainbow. In view of this scientific explanation, how can one understand the words of Scripture regarding the divine origin of the rainbow?

These verses do not state that there had never been a rainbow before the Flood. Rather, the verses state that God designated the rainbow as His sign that as long as this natural phenomenon occurs, He will not destroy the world through a flood. (I might say to a friend, "The Sun rising in the East is my sign that you can always rely on my help." These words do not mean that I am now creating the Sun. Rather, they mean that I will *always* help my friend, just as the Sun *always* rises in the East.)

Why did God choose the rainbow as His sign to mankind that another such destructive flood will never occur again? Why did He not choose a different natural phenomenon? Answering this question is the subject of this chapter.

The traditional answer is the following. Rain is a common forerunner of a flood. Therefore, when it rains very heavily, people might fear that the heavy rain is signaling the onset of another massive flood that will again destroy everything. The rainbow that accompanies the rain will serve as a sign of God's promise that a massively destructive flood will never occur again.

Elementary Particles

Many statements in Scripture have multiple explanations, a simple explanation that appeals to everyone and also a deeper explanation that is often based on science. Here, we will propose a deeper reason, based on modern science, for God's choice of the rainbow as the visual symbol of His promise not to destroy the world through a flood. Science has shown that the rainbow has a unique feature that makes it the ideal choice to symbolize this promise.

If one were to choose some object to symbolize permanence, one might think of a mountain. Nothing seems more stable than a towering mountain. However, this is a delusion. The engineer of today readily drills through a mountain to make a tunnel for a highway. If the need arises, giant earthmoving equipment can level entire mountains. Nowadays, engineers and scientists can take any object apart. Therefore, an object that symbolizes permanence should not be composed of smaller particles.

It was previously thought that the atom was an object that cannot be split into smaller particles. Indeed, the very word "atom" indicates this. The "a" means "not," as in "apolitical" or "asexual." The letters "tom" mean "cutting" from the Greek *tomos*. Thus, "atom" means something that cannot be cut into smaller pieces.

In the nineteenth century, scientists discovered that the atom *can be cut* into smaller particles. These smaller particles are the electron, proton, and neutron that make up the atom. And it has since been learned that both the proton and the neutron consist of even smaller particles, called quarks. However, there do exist particles that cannot be split into smaller particles. Scientists call them "elementary particles." They include the electron, quark, neutrino, photon, gluon, and graviton. There are other elementary particles, but they are unstable. Of course, only a *stable* particle can be chosen to symbolize *stability*. The six particles listed above are both stable and they cannot be split into smaller particles.

Thus, there are six stable elementary particles that could be used to symbolize stability and permanence. However, five of the six stable elementary particles listed above can only be detected by means of sophisticated scientific instruments. Therefore, they are not suitable to serve as God's sign to mankind, which, of course, has to be easily observable by human beings. The only stable elementary particle that can be detected without the use of scientific instruments is the *photon, the particle of light.*

The Photon

Throughout the nineteenth century, it was thought that light is a wave phenomenon, and scientists spoke of light waves. It was Max Planck who proposed in 1900 the radical idea that light actually consists of a stream of particles, which he called "photons," from the Greek word for "light." Planck's idea marked the beginning of the quantum theory, the theory used today by scientists to explain almost all natural phenomena.

Planck showed that the concept of photons — particles of light — can explain some very puzzling features of light. The idea that light consists of photons was elaborated and clarified by Albert Einstein in 1905. Both Planck and Einstein were awarded the Nobel Prize in Physics for their important contributions regarding the nature of light.

As stated previously, photons *can* be detected by the human eye without the need for scientific instruments. In fact, the human eye is such an extremely sensitive organ that the eye can detect a mere ten photons. Thus, light would seem to be the ideal choice to serve as God's sign of stability, His promise that He will never again destroy all life through a flood. But why a rainbow? Why not sunlight?

Sunlight

The color of light corresponds to the energy of the photon. Red light consists of lower-energy photons, whereas blue light consists of higher-energy photons. There are even higher-energy photons that the human eye cannot see, such as ultraviolet light and X-rays. And there are also lower-energy photons that the eye cannot see, such as infrared light and radio waves. Clearly, God's sign for mankind must consist of photons having a color/energy that the human eye can perceive.

The observed color of sunlight is nearly white. The whiteness of sunlight is due to the fact that sunlight is a *mixture* of the light of many colors, ranging from red to blue, and the eye perceives this mixture as being white.

The color of sunlight would change if photons of certain colors were removed. This actually happens at sunset, when the sunlight becomes markedly reddish-orange. The beautiful sunsets we observe are due to the fact that most of the blue photons have been removed from the sunlight, leaving reds, oranges, and yellows. At noon, sunlight has to pass a layer of atmosphere to reach the observer. The atmosphere scatters some of the photons of the sunlight, and the color of sunlight that one observes is the mixture of the colors of the remaining photons. At sunset, when the sun is close to the horizon, sunlight has to pass through a much thicker layer of atmosphere to reach the observer. This causes most of the blue photons to be scattered out of the sunlight. (For technical reasons connected with Thomson scattering, blue light is scattered more strongly by the atmosphere than other colors.) For this reason, the observed color of the sun changes during the day, being yellowish at noon and reddish at sunset.

Because the color of sunlight *changes* during the course of the day, sunlight is unsuitable to serve as a sign of *stability*.

The Rainbow and God's Covenant

Is there any occasion when sunlight consists of a single color, rather than a mixture of colors? The answer is: yes! This occurs in a rainbow. As sunlight passes through the raindrops, they act as prisms that separate the sunlight into its various colors. Each of the separated colors of the rainbow is a *single* color, consisting of *identical* photons whose color never changes. The rainbow is thus a visible symbol of the stability found in nature. Thus, one can appreciate why the rainbow was chosen as the sign of God's covenant.

> *"I hereby establish My covenant with you, with all future generations, and with all living creatures that are with you, including birds, cattle, and all the animals on Earth that were with you in the Ark. I am establishing my covenant with you that never again will all creatures be drowned by floodwaters. A flood that destroys the Earth will never again occur"* (Genesis 9:9–11).

The above verses indicate that this covenant was not restricted to the Israelites, as were later covenants (Genesis 15:18, 17:11, Exodus

24:8, 31:16, 34:10, Numbers 25:13, Deuteronomy 5:2). This covenant was not even restricted to mankind. This first covenant promised by God includes *all living creatures.*

The resplendent rainbow of photons will forever serve as God's visible sign of the constancy of nature.

Chapter 17

How God Intervenes in Our Lives Without Violating the Laws of Nature: Quantum Theory and the Butterfly Effect

Statement of the Problem

In numerous passages, Scripture clearly states that God continues to intervene in His world after the creation. There are many verses in Scripture that discuss divine reward and divine punishment. For example, Deuteronomy discusses reward (26:3–13) and punishment (26:14–44) that depend on the actions of the individual. Moreover, we are instructed to pray to God to supply our daily needs (sustenance, health, etc.). Our prayers imply that we believe that God is the ultimate source and provider of our needs. Thus, it is clear that according to Scripture, God is continuously intervening in the world.

When God created the world, He also created the laws of nature. The existence of such laws is by no means obvious. Albert Einstein said that the most incomprehensible feature of the universe is that it is comprehensible.

Jewish tradition asserts that the world operates according to well-defined laws of nature, an idea stated in the Talmud (Tractate Avoda Zara 54b). However, this does not imply that miracles — deviations

159

from the laws of nature — have never occurred. In fact, it is a basic principle of our belief that miracles have occurred. Moses Maimonides (twelfth century) wrote that one who does not believe in miracles is a heretic (*Guide for the Perplexed* 2:25). However, God's overt miracles (the Ten Plagues in Egypt and the fall of the walls of Jericho) are few and far between. We shall here discuss the manner in which God intervenes in the world on a daily basis in order to supply the needs of mankind.

What is the mechanism for this divine intervention? Does divine intervention in the world require that God alters the laws of nature in order to answer our prayers for our needs and to reward or punish according to one's actions? When we pray to God, do we expect Him to alter the laws of nature for our benefit? Since divine intervention occurs all the time, does this mean that God is constantly violating His own laws of nature?

We will discuss what science has to contribute to these questions. Our goal is to explain, on the basis of science, how God intervenes in the outcome of everyday events (e.g., providing for our daily needs) *without violating the laws of nature*. We will see that, as surprising as it sounds, God's continuous intervention need not require any deviation from the laws of nature.

To justify this statement, we must undertake a journey into the world of physics. In particular, I will discuss the two major discoveries in physics in the twentieth century: quantum theory and the theory of chaos. (No previous knowledge of physics is assumed.) These two theories pave the way for understanding the manner in which *God interacts with His world in a manner that maintains the integrity of His laws of nature*.

Classical Physics

We begin with a discussion of "classical physics," which means what was known in physics before the twentieth century.

The universe consists of objects that range in size from the minute (single atoms) to the enormous (planets and stars). One of the main tasks of physics is to explain how objects move. Explaining planetary motion was one of the basic problems in physics throughout the Middle Ages. The planetary orbits were finally accurately described in 1609 by Kepler's laws, which state that each planet revolves around the sun in an elliptical orbit, sweeping out equal areas in equal times. However, no one could explain *why* the planets move in this fashion. This problem was finally solved by Isaac Newton in 1687 in his famous book, *Principia*, one of the most important books of physics ever written. Newton formulated the law of motion, $F = Ma$. This law, known as classical mechanics, states that if force F is applied on an object of mass M, then the object will move with acceleration a.

According to classical mechanics, in order to predict how an object will move, one must know the force that is exerted on the object. Newton answered this question with regard to the planets by formulating the law of gravity. This law of nature describes the force that the sun exerts on each planet. By combining his law of motion with his law of gravity, Newton fully explained all aspects of planetary motion. Moreover, these same laws explain many terrestrial phenomena as well, such as the tides.

Although great progress in physics had been made, important problems remained, such as explaining the electrical and magnetic forces, and the phenomena of light, heat, sound, and thermodynamics.

In the 1860s, James Clerk Maxwell demonstrated that the electrical force and the magnetic force are really two aspects of the same force, now called the electromagnetic force. Maxwell also showed that light is due to this same electromagnetic force. In other words, the electrical force, the magnetic force, and light are all different aspects of the same phenomenon. This was an enormous advance in our knowledge of physics.

It was also discovered during the nineteenth century that both sound and heat are the result of molecular motion. If the molecular motion is coherent (molecules moving together), sound is produced, but if the molecular motion is incoherent (molecules moving randomly), heat is produced. The laws of thermodynamics were also established, including the law of the conservation of energy.

The progress of physics had been so rapid and so comprehensive that by the end of the nineteenth century, it was widely believed that all the basic laws of physics were now known. Consider the statement by Lord Kelvin (William Thomson), one of Britain's greatest scientists, in his 1900 address to the British Association for the Advancement of Science. Lord Kelvin declared the following:

> *"There is nothing new to be discovered in physics now. All that remains is to make more precise measurements."*

Given all the progress achieved by the end of the nineteenth century, Lord Kelvin seemed justified in saying that there was nothing fundamental left to discover in physics.

Determinism

According to classical physics, the future occurrence of every event can be predicted by the laws of physics. This is known as *determinism.* A useful way to express this idea is that *the future is already determined in the present.* In other words, if one knows the present situation of a system, then the laws of physics will determine the entire future behavior of that system. In practice, technical difficulties generally prevent one from calculating the future behavior of complicated systems. But even though one cannot predict the future due to technical difficulties, the future has already been determined in the present.

Problems with Classical Physics

Near the end of the nineteenth century, several experiments raised serious questions regarding the validity of classical physics. Two examples follow.

Radiation

When a metal is heated, it radiates energy. At sufficiently high temperatures, the metal radiates so much energy that it begins to glow. As the temperature continues to increase, the metal first glows red, then blue, and finally white.

Near the end of the nineteenth century, physicists used classical physics to calculate the amount of energy that a metal radiates as a function of its temperature. However, these calculations disagreed with the measurements. In fact, according to classical physics, *every metal should radiate an infinite amount of energy at every temperature!* This result is absurd. Something was clearly wrong with the classical laws of physics.

Structure of the Atom

Early in the twentieth century, experiments established the structure of the atom. The experiments demonstrated that an atom consists of a small central core (*nucleus*) containing positively charged particles (*protons*), which is surrounded by negatively charged particles (*electrons*) that revolve around the nucleus. The atom is thus quite similar to a miniature solar system, with the nucleus playing the role of the sun and the electrons playing the role of the planets. In the same way as the force of gravity holds the planets in their orbits around the sun, the electrical force holds the electrons in their orbits around the nucleus.

However, there is a very serious problem with this model of the atom. The electrons that revolve around the nucleus of the atom are negatively charged particles. But according to the laws of classical physics, when a charged particle moves in a circle, it radiates energy. If the electrons were to radiate energy as they circle the nucleus, they would move closer to the nucleus, eventually spiraling into the nucleus itself. If all the electrons coalesced with the nucleus, the atom would collapse. *According to the laws of classical physics, every atom should collapse in less than a billionth of a second.* However, atoms are perfectly stable and do not collapse. Therefore, something was clearly wrong with the laws of classical physics.

Quantum Theory

To solve these, and many other problems, quantum theory was formulated. It took about 30 years, from 1900 until about 1930, for quantum theory to become fully developed. Nobel Prizes were awarded left and right as the different pieces of this strange theory were discovered during the course of several decades. There was a lot of confusion regarding quantum theory in its early stages because the theory was so unexpected.

What Is the Nature of Light — Waves or Particles?

Physicists have long debated about the nature of light. Does light consist of a series of waves (as Christiaan Huygens wrote in the seventeenth century) or does light consist of a stream of particles (as Isaac Newton wrote)? Finally, in 1801, the British scientist Thomas Young demonstrated that light has wave-like properties. During the following decades, many additional experimental findings confirmed these results. By the end of the nineteenth century, no physicist doubted that light consists of a series of waves.

Quantum theory began in 1900 with the radical proposal of Max Planck that light is not a wave phenomenon after all. Planck found that he could explain the puzzling radiation of heated metals (described above) by assuming that the radiated light consists of a stream of small particles, that he called "*photons*" (from the Greek word for "light"). For his explanation, Max Planck was awarded the Nobel Prize. The word *quantum* is often used to refer to a small particle, and hence Planck's theory became known as the *quantum theory*.

Although Planck's proposal that light consists of a stream of particles explained the radiation of metals, his proposal contradicted much of the other experimental evidence that seemed to show that light consists of a series of waves.

Another experiment involving light that could not be explained in terms of waves was the photoelectric effect (the details of which need not concern us here). In 1905, Albert Einstein showed that the photoelectric effect can be explained if one adopts Planck's proposal that light consists of a stream of small particles (photons). For his explanation, Einstein was also awarded the Noble Prize.

To summarize, some experiments involving light cannot be explained unless one assumes that light consists of a series of waves, whereas other experiments involving light cannot be explained unless one assumes that light consists of a stream of particles.

This behavior of light came to be called "wave–particle duality." No one could explain this paradoxical behavior of light.

It was discovered in the 1920s that electrons also exhibited wave–particle duality, acting at times as particles, but at other times as waves. The paradox thus became more widespread.

Quantum Theory and Probability: The Future Is Not Determined in the Present

We now describe one of the strangest features of quantum theory, the feature that is directly relevant to the manner in which God intervenes in the world.

Unlike classical physics, quantum theory is *probabilistic. That means that before performing a measurement, one can never know what the result of the measurement will be.* One can only know *the probability* of obtaining each of the various possible results.

That is to say, *the future is not determined in the present.* This principle is the *exact opposite of the determinism* of classical physics.

For example, according to quantum theory, the calculation may show that the possible results of a measurement are F, G, and H, with no possibility of obtaining M or N. The calculation also shows that upon performing the measurement, there is, say, a 15% chance of obtaining F, a 45% chance of obtaining G, a 40% chance of obtaining H, and, of course, 0% chance of obtaining M or N. But this does not tell us *which of the three possible results (F or G or H)* will be obtained upon performing a measurement. *In fact, this information does not even exist.* This is the famous "uncertainty principle," for which Werner Heisenberg was awarded the Nobel Prize. In other words, there is *no meaning to the question* of whether the measurement will result in F or G or H.

It is worth emphasizing this important point. Until the measurement is actually performed, the result of the measurement remains *uncertain.* Therefore, there is no violation of the laws of nature *regardless of whether the measurement yields F or G or H.*

In complete contrast to the above scenario, classical physics asserts that there is *only one possible result* for every measurement. For the measurement under discussion, the result must be *F or G or H.* According to classical physics, which of these three results will be obtained has already been determined even before one performs the measurement.

Radium As An Example of the Probability Aspect of Quantum Theory

We begin with a bit of background. Thorium is a radioactive element, which means that every atom of thorium will eventually decay to form a different chemical element. When thorium decays, the new element formed is radium, which is also radioactive. Every atom of radium will eventually decay to form a different chemical element, which is radon.

Now we come to the point. How much time elapses before a radium atom decays to become an atom of radon? The answer is that *no one knows.* Why doesn't anyone know? One has only to place radium atoms on a table and observe how much time passes before the atoms decay. However, if one carries out this procedure, one obtains a very strange result. *The time for each radium atom to decay is different.* One radium atom may decay after five minutes, whereas another radium atom may not decay for a thousand years. And this is the case even though *all radium atoms are identical.*

According to classical physics, this result is completely impossible, for the following reason. Since all radium atoms are identical, they must all behave in the same way. This means that the time for decay for every radium atom must be the same. But according to quantum theory, in spite of the fact that all radium atoms are identical, the decay time is different for each atom. The reason is that *the future (when the radium atom decays) is not determined in the present (now).* This is the most surprising result of quantum theory.

When Is Quantum Theory Important?

The reader may be wondering why this dramatic phenomenon — *the future is not determined in the present* — had not been observed earlier. In fact, our everyday experience teaches us *just the opposite.* Throughout our lives, we observe that the future *is indeed* determined in the present. Consider a free throw in basketball, meaning that in certain circumstances, a player has the opportunity to try to throw the ball into the basket without any interference from the opposing players. Every basketball player knows that in a free throw, if the ball is thrown precisely in the direction of the basket (the present), in a few seconds (the future), the ball will enter the basket to the applause of the crowd. Why do athletes, as well as the rest of us, remain unaware of quantum theory in our daily lives?

The explanation is that the effects of quantum theory are significant *only* when describing the behavior of very minute particles, on the scale of atoms and molecules. However, when dealing with macroscopic objects, such as basketballs, the difference between the prediction of quantum theory and that of classical physics is extremely small. When a basketball is thrown in the right direction, classical physics predicts that the ball will enter the basket with 100% certainty, whereas quantum theory predicts that the chances of the ball entering the basket are 99.9999999...%, with only an extremely small chance of the ball missing the basket. Since the difference between these two predictions is immeasurably small, an athlete need not be aware of quantum theory to become a basketball star.

In everyday life, we are rarely aware of atoms and molecules. We normally deal with large macroscopic objects. Therefore, it might seem that quantum theory has no effect on our daily lives. Surprisingly, this is not the case. In fact, as we shall see, the effects of quantum theory on molecular-sized objects are enormously important.

Chaos Theory

The other scientific topic that we have to discuss is chaos theory. Chaos theory is a new branch of science, developed in 1961 by Edward Lorenz, and is widely considered to be one of the most important scientific discoveries of the twentieth century. According to *Scientific American*, "*Chaos theory represents a revolution that affects many different branches of science.*"

The concept of chaos should *not* be understood as being synonymous with confusion or mess. Quite the contrary! Chaos theory has definite rules that lead to well understood and quite surprising consequences.

The details of chaos theory are complicated, but it is easy to state the central idea. *A chaotic system is extremely sensitive to even the tiniest changes in its surroundings.* Many important systems have proved to be chaotic, including the weather.

The Butterfly Effect

The sensitivity of the weather to extremely minute changes in the atmosphere is called the "butterfly effect" ("the influence of the butterfly"). This term means that *it is literally true that a single butterfly fluttering its wings in Tokyo can significantly affect the weather in Jerusalem within about two weeks.* In other words, *extremely small changes in the atmosphere anywhere on Earth will eventually produce a significant effect on the weather everywhere on our planet.*

The butterfly effect does *not* imply that butterflies can cause unusual weather conditions, such as snow falling in Tel Aviv in August, because it never snows in Tel Aviv in August. However, a January day in Tel Aviv can be either rainy or sunny. A single butterfly fluttering its wings in Tokyo can influence which of these two scenarios will occur.

The butterfly effect is, of course, not limited to butterflies. The motion of even a small number of air molecules, as happens when a butterfly flutters its wings, can have a significant effect on the weather. And this is the case for many other systems as well. That is the great importance of chaos.

Quantum Theory, Chaos, and God's Interaction with the World

Before returning to our major theme, it is useful to summarize these two principal results.

- **Chaos theory** tells us that the motion of a relatively small number of molecules can have a *major effect* on a system, and many systems have been found to be chaotic.
- **Quantum theory** tells us that the detailed future behavior of very small systems is both *unknowable and uncertain.* Several different future behaviors of a system are possible *within the laws of nature.*

Chaos theory and quantum theory form the key points to our discussion. These two theories demonstrate how God's intervention in the affairs of mankind need not imply a deviation from the laws of nature.

Examples

Hurricanes

Hurricanes are usually generated over the ocean and their ferocity increases as they approach land. If the hurricane strikes land, it can cause enormous damage and loss of life. But if the hurricane dissipates over the ocean, it will not cause any damage at all.

Hurricanes, like all weather phenomena, are chaotic systems. Therefore, the path of a hurricane can be influenced by a relatively small number of air molecules moving in a specific direction — the butterfly effect. However, the detailed path of a hurricane cannot be calculated. According to quantum theory, *the future paths of the air molecules are not determined in the present. Various paths of a hurricane — benign or tragic — are possible within the laws of nature.* Therefore, God need not violate the laws of nature in order to grant our prayers that the hurricane will pass without causing any damage.

Coronavirus Infections

When a person infected with the coronavirus sneezes or coughs, he disperses his viruses into the air around him (if he is not wearing a face mask). Will the air molecules transmit the deadly viruses to a nearby person who is not wearing a mask? Whether the nearby person contracts corona depends on the detailed motion of the relevant air molecules. According to the principles of quantum theory, *it is unknown and unknowable* whether or not the air molecules will move in such a way to cause the virus to reach a nearby unmasked person who will then become infected. *Both scenarios are possible within the laws of nature.* Therefore, God need not violate laws of nature in order to grant our prayers for health.

Accidents

Older people may lose their balance and fall. One older person may break his hip and be crippled for life, whereas another older person may receive nothing more serious than a bruise that soon heals. It all depends on the minutest details of how the person fell.

However, the principles of quantum theory assert that such minute details are *unknown and unknowable. Within the laws of nature,* either a safe or a debilitating outcome of a fall is possible. Therefore,

God need not violate the laws of nature to grant a safe outcome to a fall.

Conclusion

We have seen how the startling predictions of quantum theory and of chaos theory pave the way for understanding how God may continuously intervene in the affairs of mankind *without violating His laws of nature.*

Chapter 18

Scientific Blunders

Anyone who deals with the subject of Science and Scripture will, sooner or later, be confronted with the question of fossils, those ancient relics of prehistoric plants, animals, and men. The common assumption is, of course, that the scientific evidence regarding fossils is basically reliable and need not be seriously questioned by the layman. In other words, one's job is *to deal* with these facts — not to cast doubt on their veracity. Other than the creationists, who reject the entire scientific enterprise, one naturally assumes that the fossil evidence and its interpretation have been presented by serious scientists who were objective in their pursuit of knowledge and who used accepted standards of scientific rigor.

In this chapter, we shall see that, unfortunately, this has not always been the case. In fact, blunder after blunder has been made in the course of "scientific work" regarding some fossils. And the blunders regarding these fossils were carried out relatively recently, in the twentieth century. Moreover, the absurd interpretations of these fossils were *not* published by third-rate workers, but *by world-famous scientists*. Indeed, it is precisely because of the unquestioned authority of these scientists that it often required decades to correct their errors. It has become clear in recent years that the reason for this shoddy work is that the evolutionists in question were, unfortunately, often motivated by subjective considerations, including national pride, professional jealousy, and preconceived notions.

Only *leading authorities in evolutionary biology* will be quoted here. This is an important point, because in recent years, a plethora of books and articles criticizing Darwin and evolution has been published, written by lawyers, journalists, physicians, and other laymen in the field of evolutionary biology. It should be emphasized that such books and articles will *not* be quoted here. The sources presented here were written by experts in evolutionary biology.

Hesperopithecus: The Man Who Was a Pig

The fossil that we shall discuss bears the scientific designation *Hesperopithecus* ("western ape-man") to emphasize that this claimed fossil of prehistoric man was discovered in the Western Hemisphere. It will be the rare reader who has ever heard of *Hesperopithecus*. The history of this fossil has been shoved deep under the rug by almost all paleontologists, and its story has been carefully excised from all scientific writings — and with good reason. Of all the blunders committed by evolutionary biologists during the twentieth century in their various "scientific studies" of man-like fossils, none can compare with *Hesperopithecus*.

Our story takes place in America in the 1920s, a decade marked by an ongoing battle between creationists and scientists. Today, creationists make much more modest demands. They merely insist that their views be taught in the schools side by side with standard evolutionary theory. In the 1920s, however, their demands were considerably more far-reaching. They insisted that *only* their ideas be taught. In fact, the creationists succeeded in passing laws in several states, including Tennessee, making it a criminal offense to teach Darwin's theory of evolution in the public schools. The opposition convinced a high-school teacher in Tennessee named John Scopes to openly teach Darwin's theory in order to challenge the law.

The resulting Scopes trial became a national sensation, pitting the foremost trial lawyer of the day, Clarence Darrow, against one of America's leading creationists, William Jennings Bryan (who was nearly elected President of the United States!). It is necessary to understand this background and the mood of the country to properly appreciate the story of *Hesperopithecus* and its impact on scientific thinking.

The "hero" of the *Hesperopithecus* fiasco was Henry Fairfield Osborn, a leading evolutionary biologist. Unfortunately, Osborn was often motivated by personal rivalry and preconceived notions, features which paved the way for a scientific disaster. Osborn was universally recognized as "a great paleontologist"[1] and served as the director of the world-famous American Museum of Natural History in New York.

His feelings for Bryan have been described[2] as "pure venom and contempt. In Osborn's view, Bryan was perverting both science and the highest notions of divinity." Osborn hated Bryan with a passion, setting the stage for a vicious confrontation between the two men in the arena of "Science and Scripture."

The Osborn–Bryan confrontation began in February 1922, when the *New York Times* published an article by Bryan attacking Darwin's theory of evolution, soon followed by a reply from Osborn which defended the scientific principles of evolution and argued that the concept of evolution was, in fact, completely compatible with Scripture. To ridicule Bryan's rejection of fossil evidence, Osborn cited a passage in Scripture: "*Speak to the earth and it shall teach you*" (Job 12:8). Osborn was referring, of course, to the fossil evidence.

In the month following this sharp exchange in the *New York Times*, a geologist sent Osborn a fossil tooth that he had discovered. Osborn created a sensation with the claim that this was the first fossil of prehistoric man ever found in America. This claim was to lead to Osborn's downfall. Stephen Jay Gould has described what happened[3]:

---

[1] S. J. Gould, 1991, *Bully for Brontosaurus* (Penguin Books: London), pp. 432–447.

[2] S. J. Gould, 1991, *Bully for Brontosaurus* (Penguin Books: London), p. 433.

[3] S. J. Gould, 1991, *Bully for Brontosaurus* (Penguin Books: London), pp. 434–436.

"Osborn's enthusiasm warmed as he studied the tooth and considered its implications. An American prehistoric man would certainly be a coup for Osborn's argument that the earth spoke to Bryan in the language of evolution.

Therefore, Osborn proclaimed the momentous first discovery of a direct human ancestor in America. Osborn named the fossil Hesperopithecus and presented it to the scientific world in a paper published in the April 1922 issue of the prestigious Proceedings of the National Academy of Sciences.*"*

Osborn exulted in the uncannily happy coincidences of both time and place. Not only was this fossil discovered at the very time that Bryan was denying fossil evidence. The crowning irony was that *Hesperopithecus* was found in Nebraska, Bryan's home state! No fossil could have had a greater potential to embarrass Bryan. No fossil could have bettered *Hesperopithecus* for rhetorical impact. Needless to say, the precious irony of the situation was not lost on Osborn, who inserted the following gloat of triumph into his article for the staid *Proceedings of the National Academy of Sciences*:

"It has been suggested humorously that the fossil should be named Bryanopithecus after the most distinguished primate that the State of Nebraska has ever produced. It is certainly ironic that I had advised William Jennings Bryan to consult a certain passage in the Book of Job, 'Speak to the earth and it shall teach you,' *and it is a remarkable coincidence that the first earth to speak on this subject is the sandy earth of Snake Creek in western Nebraska. "*

For several years after the discovery of *Hesperopithecus* in 1922, Osborn missed no opportunity to use the fossil to heap public abuse on Bryan. On the eve of the Scopes trial in 1925, Osborn published a book devoted primarily to ridiculing Bryan and chose a biting parody of Job as his title: *The Earth Speaks to Bryan.*

In addition to using *Hesperopithecus* to attack Bryan, Osborn publicized the prize fossil in his American Museum of Natural History, commissioning a graphic reconstruction of a *Hesperopithecus* couple in a forest surrounded by other members of the Nebraska fauna, prepared by the well-known scientific artist Amadee Forestier.[4] This reconstruction was a marvelous example of the lifelike three-dimensional exhibits for which this museum is justly famous. Looking at a photograph of the well-known *Hesperopithecus* exhibit, one cannot help but be amazed by the many details of the physical appearance and the cultural behavior of this prehistoric man and woman that Osborn claimed to have deduced *from one single tooth.*

Five years later, Osborn's world collapsed. Additional fossil evidence discovered in Nebraska showed conclusively that the *Hesperopithecus* fossil was, in fact, the tooth of a *pig*. Osborn's longstanding claim that *Hesperopithecus* was a fossil of prehistoric man was officially retracted in the 16 December 1927 issue of *Science*.

As a sorry comment on Osborn's integrity, it should be noted that his name does not appear in the retraction article. He left to a younger colleague the embarrassing task of admitting publicly that their famous *Hesperopithecus* "prehistoric man" was really nothing but a pig.

How could this farce happen? Why were America's leading paleontologists so ready to accept the absurd idea that a *single tooth* — so worn that it could not even be properly identified as belonging to a pig — was sufficient to establish a new class of prehistoric men? To answer this question, one must be aware of the situation at that time in paleontology. By the 1920s, man-like fossils had been found worldwide, everywhere but in America. France had Cro-Magnon Man, Germany had Neandertal Man, Asia had Peking Man and Java Man, and Africa had *Australopithecus*. Man-like fossils were being discovered everywhere — except in the United States. (It has since become recognized that the prehistoric men entered the Western Hemisphere only about 20,000 years ago.)

American paleontologists had been relegated to being mere spectators in the prestigious game of man-like paleontology. Therefore, when the *Hesperopithecus* fossil was discovered, they eagerly jumped onto Osborn's bandwagon. In spite of his unques-

---

[4] S. J. Gould, 1991, *Bully for Brontosaurus* (Penguin Books: London), p. 444.

tioned scientific talents, objectivity was not a feature of Henry Fairfield Osborn.

As a result of this scientific fiasco, over a million visitors to the American Museum of National History in New York were enthralled by the brilliantly executed reconstruction of the *Hesperopithecus* prehistoric man and woman living in the Nebraskan forest. Few of these visitors would ever read the scientific literature which revealed the truth about "the man who was actually a pig."

David Pilbeam, of Harvard University, discussed this lamentable situation at length and made the following remarks[5]:

> *"Our theories have often said far more about the theorists than about what actually happened ... All our theories about human origins were relatively unconstrained by fossil data ... Many evolutionary schemes were dominated by theoretical assumptions that were divorced from the data derived from fossils."*

Life on Mars?

The headline of the *New York Times* screamed: "NEW HINT OF LIFE IN SPACE: Meteorites Yield Fossilized, One-Cell Organisms Unlike Any Known on the Earth." *Newsweek* asserted dramatically: "Something Out There!" Respected scientists told crowds of reporters that their research work, published in a prestigious journal, revealed complex hydrocarbons and what appeared to be fossilized bacteria buried deep within a meteorite. They claimed to have found "the first physical evidence for the existence of forms of life beyond our planet."

The events described above occurred in 1961 and the meteorite in question had fallen in Orgueil, France, more than a century earlier. However, the "fossils" ultimately proved to be ragweed pollen and the "organic chemicals" turned out to be furnace ash!

---

[5] D. Pilbeam, 1980, in *Major Trends in Evolution*, editor, L. K. Konigson (Pergamon Press: London), pp. 262, 267.

How could such blunders have occurred? One would have thought that as soon as the meteoritic substances were examined by leading authorities, the truth would *immediately* become evident. How could any scientific expert fail to distinguish between "fossilized bacteria" and ragweed pollen or between "prehistoric complex hydrocarbons" and furnace ash? But that is *exactly* what happened! Indeed, *many years* were to pass before the embarrassing truth finally became recognized. *As late as 1971,* a well-known biology textbook, written by a Harvard University professor, stated: *"Evidence for life in other parts of the cosmos came from the discovery in 1961 of what were identified as fossils of microscopic organisms, like algae, in meteorites."*[6]

History Repeats Itself

Three decades later, on 7 August 1996, history repeated itself. The story began with a dramatic announcement by NASA Administrator Daniel Gordin (NASA is the U.S. Space Agency) that "a meteorite, consisting of a chunk of Mars rock, bears evidence of ancient life."[7] The NASA research team, headed by geologist David McKay, claimed that "the peculiar features found in meteorite ALH84001 are best explained by the existence of primitive life on early Mars."[8]

Just what had been discovered? Under discussion is a meteorite — essentially a rock about the size of a grapefruit and weighing 1.9 kilograms — found in the snows of Antarctica. For technical reasons that need not concern us here, this meteorite (denoted ALH84001) is believed to have originated on Mars. The excitement centered on the possibility that living creatures had once inhabited the meteorite. No signs of life were actually found in this meteorite, but a number of different types of minerals were present in configurations that are usually produced by bacteria, the smallest and simplest living cells. On this basis, the scientific article, written by NASA researcher McKay and his colleagues, exclaimed: "When

---

[6] C. A. Villee, 1971, *Biological Principles and Processes* (Saunders: Philadelphia), p. 320.

[7] *News, Science,* 16 August 1996, vol. 273, p. 865.

[8] News and Analysis, October 1996, *Scientific American,* p. 12.

these phenomena are considered collectively, in view of their spatial association, we conclude that they are evidence for primitive life on early Mars."[9]

The bubble burst before the end of the year. The December 1996 issue of *Scientific American* carried the following news item:

> *"Believers had a big thrill last summer when NASA announced that they had uncovered signs of Martian life in a meteorite. The evidence came in the form of tiny, sausage-shaped imprints, which the scientists said were most likely left by nanobacteria."*

However, researchers at the Massachusetts Institute of Technology subsequently demonstrated that purely inorganic happenings can make identical marks.

A British science journal was even more dramatic in presenting the evidence against life on Mars. Its news item, sarcastically entitled "Death Knell for Martian Life — New Studies Indicate That This 'Fossil' Was Never Alive," begins as follows[10]:

> *"As 1996 draws to a close, those heady days of summer seem like a dream. In August, scientists led by NASA's David McKay stunned the world by producing evidence of past life on Mars. But now, two new analyses put the final nail in the coffin of that claim."*

Objectivity and Budgets

How could the embarrassments of 1961 be repeated so soon? Perhaps some insight can be gained by examining the immediate effects of the claims by NASA scientists of having found evidence for life on Mars.[11]

> *"The NASA announcement generated an almost unheard-of agreement between House Speaker Newt Gingrich and Vice-President Albert Gore for*

---

[9]D. S. McKay *et al.*, 16 August 1996, *Science*, vol. 273, pp. 922–930.

[10]*This Week*, 21 December 1996, *New Scientist*, vol. 150, p. 4.

[11]*News*, *Science*, 16 August 1996, vol. 273, p. 865.

additional government spending ... President Bill Clinton himself announced that Gore will organize a White House meeting to map out a bipartisan course to focus the U.S. space program on the issues raised by the new findings."

This news was heaven-sent for NASA. U.S. space science had been laboring under serious budgetary constraints since the 10% cut in funding contained in President Clinton's 1997 budget request. Even worse were the NASA budget cuts scheduled for the next five years. The pessimistic words of the NASA space science advisory committee set the tone: "We see the handwriting on the wall. The outlook is bleak."[12]

Everything suddenly changed with the dramatic findings about life on Mars.

"These findings are already breathing new life into solar system exploration, with Jerry Lewis, chairman of the U.S. House of Representatives panel that oversees NASA funding, stating that he now supports increasing the agency's budget to accommodate more aggressive exploration of Mars."[13]

Similarly, John Logsdon, of the Space Policy Institute at George Washington University, pointed out that following the NASA announcement of their discovery of life on Mars, "the extreme cuts to NASA's budget floated in budget projections earlier this year now seem unlikely."[14]

Since the 1986 *Challenger* disaster (in which seven astronauts died) and other failures, NASA had come under increasing criticism regarding the running of its space program. A dramatic success, such as the discovery of life on Mars, would do much to bolster its public image. Therefore, there are valid reasons for questioning the scientific objectivity of NASA's wildly optimistic claims regarding the evidence for life on Mars.

---

[12] *News, Science*, 22 March 1996, vol. 271, p. 1660.

[13] *News, Science*, 16 August 1996, vol. 273, p. 865.

[14] *News, Nature*, 7 November 1996, vol. 384, p. 4.

Comment

The bizarre examples of scientific blunders presented in this chapter carry an important message. As a professional scientist for several decades, I greatly respect and value the scientific enterprise. However, scientific research is carried out by human beings, who are subject to prejudice, national pride, and budgetary considerations that afflict us all. The history of science has shown that subjective behavior has plagued some of the most famous scientists. This does not mean that science as a whole is suspect and should not be accepted. Science has very strong track record.

Chapter 19

The Claim of Intelligent Design

Introduction

The modern concept of Intelligent Design was proposed in 1996 by biochemist Michael Behe in his book, *Darwin's Black Box, the Biochemical Challenge to Evolution.* Behe claimed that he discovered *ironclad proof* for the existence of a supernatural being, whom he called the "Intelligent Designer." His studies of the living cell led Behe to conclude that Darwinian evolution cannot explain many of the biochemical reactions that take place in the cell, and that only Intelligent Design can explain them. Although Behe refrained from identifying the Intelligent Designer, the widespread understanding is that the Intelligent Designer is God.

In this chapter, we shall explain what Behe claimed and why his claim is wrong.

Important Preliminary Remarks

Before continuing, it is important to make two preliminary remarks.

- The question before us is *not* whether or not a supernatural being is ultimately responsible for the many biochemical reactions that take place in all living creatures. As a person of faith, I certainly do believe in a Supreme Being — God — Who created the world, together with the laws of nature, and God is responsible for the biochemical systems that exist in the cells of every living creature.

The question before us is the following: *Has Behe* **proved** *that this is the case or does belief in a Supreme Being remain a matter of* **faith**?

- The question before us is not whether God is the source of the laws of nature. The answer to this question is a matter of faith. However, every person, both believer and non-believer, agrees that the laws of nature are able to explain why the sun rises every day in the East and why water boils at a temperature of 100 degrees.

The question before us is the following: *Are the laws of nature also able to explain all the biochemical reactions that take place in the living cell?*

Behe claims that the laws of nature are *unable* to explain the living cell. However, H. Allen Orr, professor of evolutionary biology at the University of Rochester, has published definitive proof that Behe erred in his claim. Orr demonstrated that the laws of nature *are able* to explain the functioning of the living cell.

History of Intellent Design

Behe's proposed proof that the cell could not have been formed through Darwinian evolution has generated enormous interest (reported in *Newsweek, U.S. News & World Report, New York Times, Commentary, National Review,* and many other periodicals).

Unlike previous individuals who attacked evolution, Michael Behe is a Professor of Biochemistry at a respected university, a research scientist who performs experiments, is awarded grants, and publishes papers in international science journals. Moreover, his book is extremely well written and shows his expertise in biochemistry. Indeed, Behe's book is the most sophisticated attack on evolution to appear in recent years. However, as we shall see, Behe's proof for the existence of a supernatural entity is based on one single claim, and that claim is erroneous.

Irreducible Complexity

Behe's claim for the existence of Intelligent Design (ID) is based on what he terms "irreducible complexity." How does irreducible complexity differ from the usual forms of complexity? On what basis does Behe claim that irreducible complexity *cannot be explained* by gradual Darwinian evolution?

Darwinian evolution operates through the appearance of mutations in the genetic makeup of an animal. Sometimes, the mutation enhances the animal's chances for survival by making the animal a bit stronger or faster or less susceptible to disease, etc. Such a mutation is called a "favorable mutation." An animal with a favorable mutation is more likely to live long enough to reproduce the next generation. Therefore, a favorable mutation is likely to become incorporated into the species gene pool. The accumulation of numerous favorable mutations over many generations brings about extensive changes in the animal, eventually leading to an entirely new species.

The key point is that *only favorable mutations that enhance the animal's chances for survival* are likely to become incorporated into the gene pool.

Behe asserts that the gradual accumulation of favorable mutations *cannot* explain the development of many vital biochemical mechanisms. Among the various examples cited by Behe is the mechanism for blood clotting. Twelve biochemical reactions are involved in blood clotting, and *if even one of these twelve biochemical reactions does not occur*, the blood will not clot.

The central claim of Behe is that the complicated mechanism for blood clotting could not have evolved gradually through a series of mutations, *with each mutation providing an additional survival advantage to the animal.* Each such mutation would, by itself, be *useless.* The *complete* 12-step blood-clotting mechanism had to appear in the gene pool *all at once.* However, the probability is completely negligible that twelve specific mutations will appear in the gene pool *simultaneously.*

Refutation of ID

In December 1996, shortly after the publication of Behe's book, evolutionary biologist H. Allen Orr, of Rochester University, published an article in *Boston Review* that refuted the claim of ID. Orr demonstrated that the specific twelve mutations needed for blood clotting *could indeed appear in the gene pool gradually*, one after the other, *through intermediate stages*, with each **intermediate stage providing an additional survival advantage to the animal**. This process is, of course, in complete accord with Darwin's theory.

We present the details of Orr's refutation of Behe's claim in the Appendix.

The Situation Today Regarding ID

In view of the fact that H. Allen Orr has conclusively proved that the central thesis of ID is incorrect, one would expect that interest in ID would have died down. But a Google search shows that this is not the case. To this very day, articles are published that claim that ID is correct, even though their authors do not seem to be aware of the basis for Michal Behe's claim for the validity of ID. Similarly, articles are published that denounce ID, even though their authors do not seem to be aware of the proof published by Orr showing that ID is false. The senseless arguments continue.

Appendix

Here, we present Orr's explanation of how the twelve mutations necessary for blood clotting could have *entered the gene pool gradually, through intermediate stages, with each intermediate stage providing an additional survival advantage.* And this is the case even though *all twelve mutations are necessary* for the 12-step biochemical reaction that causes the blood to clot.

The following table lists the influence of each mutation on the biochemical system.

Mutation number	Biochemical system	Improvement?	Irreducible system?
0	**A**	–	–
1	**A + B**	yes	No
2	**A* + B**	yes	yes
3	**A* + B + C**	yes	No
4	**A* + B* + C**	yes	Yes
–	–	–	–
22	**A* + B* + C* + D* + E* + F* + G* + H* + I* + J* + K* + L**	Yes	yes

The explanation of the Table is the following. In the distant past, a biochemical system may have consisted of only one part, say part **A**. The system worked, although not too well. A genetic mutation then produces part **B**, which leads to a somewhat improved system, consisting of **A + B**. This improved system is *not* irreducibly complex, because the improved system will function without part **B**. A second genetic mutation then transforms **A** into **A***, which leads to a further improvement of the system. However — *and this is the crucial point* — **A*** will not work unless **B** is present. Therefore, the system consisting of **A* + B**, *is* irreducibly complex because *both **A*** and **B** are necessary for the system to function.*

We have thus shown how an irreducibly complex system can be produced by gradual evolution, with each mutation leading to an improvement in the system, even though the final system (**A*** + **B**) *will not function at all* unless both its parts are present. Therefore, we are done. The claim of ID — *that this is impossible* — has been refuted.

Let's continue. A third genetic mutation now occurs to produce part **C**, which leads to further improvement. This system is *not* irreducibly complex, because it will function without part **C**. A fourth mutation transforms **B** into **B***, yielding yet another improvement. However, **B*** will not work unless **C** is also present. Therefore, the system (**A*** + **B*** + **C**) *is* irreducibly complex because *all three parts are necessary for the system to function.* Nevertheless, this irreducibly complex system was produced *by a series of gradual improvements,* in accordance with Darwinian evolution.

This process can be continued to produce the twelve-part irreducibly complex system: **A* + B* + C* + D* + E* + F* + G* + H* + I* + J* + K* + L***.

An important feature of this procedure concerns its *irreversibility.* After the biochemical system is complete, one cannot determine the order in which its twelve parts were formed, or what were the intermediate parts (**A, B, C, D, E, F, G, H, I, J, K, L**). Once the scaffolding has been removed, there is no way to determine how the irreducibly complex building was constructed.

Chapter 20

Can Science Prove the Existence of God?

Seeking proof for the existence of God was a matter of great importance to medieval philosophers, both Jewish (e.g., Moses Maimonides) and Christian (e.g., Thomas Aquinas). Why was it so important to these outstanding thinkers to prove that God exists?

To answer this question, one must return to the period that preceded modern science. In the ancient world, discovering the laws of nature by experimentation was a foreign idea. Mathematicians had discovered the laws of geometry by pure reason, and it was thought that this was the appropriate method for studying the physical universe as well. Performing careful experiments seemed unbecoming to the philosopher. His realm of activity was the mind. Only a servant or an artisan would "get his hands dirty" with the many menial tasks required to carry out an experiment. An exception was astronomy. The ancients excelled at observing the motion of the heavenly bodies. Since the heavenly bodies were the handiwork of the Creator, observing their motion was not viewed as degrading. However, examining earthly objects was considered inappropriate for the philosopher — the thinker. Thus, we find in some ancient philosophical texts that in contrast to a man, a woman has only twenty teeth (the correct number for both sexes is thirty-two). It did not occur to the scholastic philosopher to actually count a woman's teeth. Such a prosaic act was completely unnecessary. Everything could be determined by reason, logic, and thought.

The above approach was not limited to the study of the universe. It was believed that *all* fundamental questions could be answered by logical deduction and pure reason. Since medieval theologians believed that God exists, they naturally assumed that His existence must be susceptible to rigorous proof. In their eyes, the inability to prove that God exists would cast doubt on His existence.

Because of their reverent attitude toward the power of logic, many theologians devoted considerable effort to arguments intended to prove that God exists. Although this subject is nowhere discussed in Scripture, proofs for the existence of God are a major topic in the writings of medieval theologians. It is instructive to analyze these arguments and their shortcomings.

The Prime Mover Argument

The most famous proof for the existence of God is called the "prime mover argument." (The word "argument" does not indicate a dispute. It is an old English word for "proof.") We all experience in our daily lives the truism asserted by Aristotle: "There is no motion without a mover." When I rearrange the living room furniture under the watchful eye of my wife, I am painfully aware of the fact that the heavy couch will not budge even one millimeter unless I push it, and the instant that I stop pushing, the couch ceases to move. If I throw a ball, its motion persists even after it leaves my hand because I have imparted some "impetus" to the ball. According to the widely accepted "impetus theory," the ball will continue to move until it uses up all its acquired impetus. Then, the ball will come to rest because "there is no motion without a mover."

Let us now turn our attention to the heavenly bodies, whose ceaseless motion is observed day after day, year after year, century after century. What causes the ceaseless motion of the heavenly bodies? Who is their "mover"? It can only be God. Thus, one has proven the existence of God.

The bubble burst in the seventeenth century, when Isaac Newton formulated his three laws of motion in his *Principia*, the most important book of physics ever written. Newton's first law of motion, the law of inertia, states that, in complete contrast to Aristotle, a moving body *will continue to move forever* unless some force causes it to stop moving. *One does not need a force to maintain the motion.* One needs a force to end or alter the motion. In the above example, the force that causes the furniture or the ball to stop moving is friction. However, if friction were not present, then the motion would persist forever. In heaven, there is no friction. Therefore, according to the law of inertia, heavenly bodies will move forever *without any agency being required to keep them moving.*

To complete the picture, Newton's law of inertia predicts straight-line motion, whereas the orbit of the planets is an ellipse. This is due to the gravitational attraction between the sun and the planets, which yields the observed elliptical orbits. Planetary motion is completely described by Newton's laws, *without the need to invoke a supernatural entity.* The prime mover argument for the existence of God is thus refuted.

Sequel

There is an interesting sequel to this story. As explained above, the refutation of the prime mover argument follows from Newton's scientific discoveries. Isaac Newton was a deeply religious man, for whom belief in God was of paramount importance. When he realized that his scientific findings had refuted one of the primary proofs for the existence of God, he was troubled. After giving the matter considerable thought, Newton "solved" the problem in the following ingenious way.

The path of a planet around the sun would be a perfect ellipse if the only force acting on the planet was gravitational attraction of the sun. In fact, however, each planet is also influenced by the gravitational attraction of all the other planets. Although this additional force is very small because the sun is a thousand times more massive than all the planets combined, it may nevertheless have a significant effect on planetary motion. Therefore, Newton set out to calculate the effect on planetary motion due to the gravitational attraction of the other planets. This difficult calculation, known as the many-body problem, cannot be solved exactly but only approximately. Newton introduced an approximation that he considered adequate, and calculated the effect on planetary motion. The result was startling! Newton found that the solar system was *unstable.*

The small gravitational force exerted by each planet on every other planet will disrupt the stability of the entire solar system. According to Newton's calculations, each planet will slowly drift away from the sun. Eventually, the sun will be devoid of planets and the solar system will cease to exist.

However, thousands of years of continuous star gazing had firmly established that the solar system was, in fact, stable. The planets do not move further away from the sun in the course of time. They move in stable elliptical orbits, with each planet maintaining the same maximum distance from the sun. How could the contradiction between Newton's calculations and the observations be resolved? *The explanation given by Newton was that it must be God who maintains the stability of the solar system!* It is the Supreme Being Who pushes the planets back when they tend to drift away from the sun.

Thus, Newton had found that after all, there is a need to invoke God to explain planetary motion. The prime mover argument for the existence of God, suitably modified, had been restored.

A hundred years later, the brilliant French mathematician Pierre-Simon Laplace reexamined the problem of the stability of the solar system. Laplace was able to show that Newton's approximation was inadequate. Laplace introduced a more accurate approximation, solved the resulting equations, and found that the solar system *was* stable after all. In other words, the instability of the solar system claimed by Newton did not really exist. It was an artifact of his inadequate approximation.

The conclusion to be drawn from this discussion is that the prime mover argument is erroneous. It is not possible to prove the existence of God from the motion of objects — not the heavenly bodies nor the more mundane earthly objects.

God of the Gaps

The "prime mover argument" for the existence of God was based on a lack of knowledge of physics. This argument is an example of what is called the "God of the gaps." When a phenomenon seems inexplicable, one says, "Aha! It must be God Who is causing this phenomenon." The problem with this approach is that the "inexplicable" phenomenon ("gap" in our knowledge) invariably becomes explained as science progresses.

Even if one can find no explanation for a certain phenomenon, the response of the scientist should be: "I'll think about it." The response should *not* be that since I cannot think of a scientific explanation, *it follows* that the phenomenon in question must have been caused by God.

Unfortunately, to this very day, there are people who contend that the existence of God can be proved on the basis of scientific discoveries. The universe has been shown to be very subtle and complex, and these people claim that the complexity of the universe could only have come about through the actions of God. This "proof," known as the "argument from design," has long since been shown to be erroneous. It is well to realize that belief in the existence of God is a matter of **faith**, and not a matter of **proof**.

Author Index

Subject Index